# Smart Biofeedback - Perspectives and Applications

*Edited by Edward Da-Yin Liao*

Published in London, United Kingdom

IntechOpen

*Supporting open minds since 2005*

Smart Biofeedback - Perspectives and Applications
http://dx.doi.org/10.5772/intechopen.92490
Edited by Edward Da-Yin Liao

Contributors
Jeff Tarrant, Valeska Kouzak Campos Da Paz, Aloysio Campos Da Paz Neto, Ivo Donner, Shao-I Chu, Hui-Li Wang, Jiun-Han Yen, Yu-Jung Huang, I-Yueh Fang, Wei-Chen Lin, Chao-Tien Hsu, Chih-Lung Hung, Tzung-Ching Lin, Te-Tsun Shen, Shu Ya Pan, Robert W. Thatcher, Kajal Patel, Manoj Sivan, Anthony Jones, James Henshaw, Edward Da-Yin Liao, Carl J. Biver, Ernesto Palermero Soler, Joel Lubar, J. Lucas Koberda



First published in London, United Kingdom, 2021 by IntechOpen
IntechOpen is the global imprint of INTECHOPEN LIMITED, registered in England and Wales, registration number: 11086078, 5 Princes Gate Court, London, SW7 2QJ, United Kingdom
Printed in Croatia

British Library Cataloguing-in-Publication Data
A catalogue record for this book is available from the British Library

Additional hard and PDF copies can be obtained from orders@intechopen.com

Smart Biofeedback - Perspectives and Applications
Edited by Edward Da-Yin Liao
p. cm.
Print ISBN 978-1-83881-973-6
Online ISBN 978-1-83881-978-1
eBook (PDF) ISBN 978-1-83881-979-8

For EU product safety concerns:
IN TECH d.o.o., Prolaz Marije Krucifikse Kozulić 3, 51000 Rijeka, Croatia, info@intechopen.com or visit our website at intechopen.com.

# Meet the editor

Edward Da-Yin Liao is the co-founder of Straight & Up Intelligent Innovations Group Co., San Jose, CA, U.S.A. He received his BS in Mechanical Engineering and an MS and PhD in Electrical Engineering, all from National Taiwan University, Taiwan. He is creative and business-savvy, with more than two decades of progressive experience in semiconductor and high-tech industries. He has been an executive of four international IT companies, including two in Silicon Valley, one in London, and one in Taiwan. He was also a university professor in Taiwan. Dr. Liao evangelizes deep learning, robotics, automation, the Internet of Things, next-generation wireless communications, and smart and healthy living.

# Contents

# Preface

The first time I heard the term "biofeedback" was in 1995. At that time, I was helping my wife (my girlfriend then, a DDS who was pursuing her MS in Dental Surgery) in preparing her thesis and experiments. Her study interest was about the *Efficacy of Occlusal Splints for Treating Sleep Bruxism (Tooth Grinding)*. Her experiments were to study the *Influence of Visual and Periodontal Feedbacks on the Accuracy and Sustainability of Biting Force Control*. The experiments started with development of a handmade, fork-type strain gauge force transducer whose analog outputs were AC/DC-converted to digital signals and shown in a digital display board (DDB). Each experiment was conducted in two phases: the unanesthetized phase and the anesthetized phase. There were two sections for each phase and four trials were conducted in each session. In the first session of the unanesthetized phase, an unanesthetized subject was asked to maintain biting forces of 5 and 15 kilogram-force (kgf), respectively, at the incisor and both sides of first molar regions for 15 seconds, both with visual biofeedback from the digits shown in the DDB. In the second session of the same phase, the DDB was turned off (without biofeedback) in the last 10 seconds. After completing the unanesthetized phase, the right upper and lower first molar regions were anesthetized with 2 percent lidocaine. Then the same experiment procedures were conducted again in the anesthetized phase.

The experiments were busy and tedious. During an experiment, all the numbers shown in the DDB needed to be observed, read, and recorded on paper. Results were put into a spreadsheet for further analysis using statistics tools like SPSS. All these ETL (extract, transform, load) operations were done manually. For the 15-second test period, only three to five data points were read and written down. It's in Year 1995, the Age of Pentium 486 and Windows 95—a Stone age, from today's ICT (information and communications technology) point of view.

Now is the 2020s. The emerging technologies and applications have shaped the future of many technologies like smart biofeedback.

This book provides readers with information on research projects and practices in modern smart biofeedback. It is designed for medical specialists, clinical practitioners, researchers, developers of biofeedback technologies, and scholars and graduate students in the fields of biomedical engineering, medicine, physiotherapy, computer engineering, data science, and so on. Chapters cover topics such as brain networks, neuromeditation, psychophysiological psychotherapy, physiotherapy, and privacy, security, and integrity of data.

I am grateful to the chapter authors for their contributions. I especially wish to acknowledge their patience in putting up with repeated requests for revision amid the COVID-19 pandemic. I would also like to acknowledge Ms. Ana Simcic, Ms. Dajana Pemac, and Anke Beck from IntechOpen for their enthusiasm and support for this book from conception to completion.

November, 2020

**Edward Da-Yin Liao**
Straight & Up Intelligent Innovations Group Co.,
United States of America

## Section 1

# Introduction

Chapter 1

# Introductory Chapter: Smart Biofeedback – Perspectives and Applications

*Edward Da-Yin Liao*

## 1. About this book

This book is about the *perspectives and applications in smart biofeedback*. The chapters of this book provide a glimpse of the research projects and clinical applications that are underway in smart biofeedback as well as the perspectives of smart biofeedback.

Biofeedback is an autonomic feedback mechanism to let people (1) observe their physiologic information such as muscle tension, blood pressure, heartbeat rates, and brain wave signals, (2) develop awareness of their physiological reactions, and (3) learn to change these physiologic responses accordingly. The biofeedback mechanism helps people take responsibility for their cognitive, emotional, and behavioral changes for better health. A biofeedback process may utilize many and various sensors to measure physiologic functions and parameters. Sensory data are first collected, analyzed, and then fed back to the human sensor nervous system in a simple, direct and immediate way.

Modern biofeedback experiments began in the early 70's of the last century [1–3], with a promising evidence observed on the efficacy of clinical biofeedback applications [4–7]. For past five decades, uses of biofeedback techniques have been widely seen in health, wellness, awareness, and computer interactive entertainment, including therapies of self regulation of psychiatric and physiologic disorders [8–21], rehabilitation [22–28], healthcare [29–31], training activities such as balancing [32, 33], postural [34, 35], relaxation [36] and sports [37, 38], development of socio-emotional interactions [39, 40], assessment of psychophysiological stress [41–44], falling prevention and detection [45], computer game design [46], and more.

Smart biofeedback [47–52] is receiving attentions because of the widespread, available building blocks of advanced technologies and smart devices that are used in effective collection, analysis and feedback of physiologic data. Researchers and practitioners have been working on various aspects of smart biofeedback methodologies and applications by using wireless communications [53, 54], Internet of Things (IoT) [55, 56], wearables [57, 58], biomedical sensors [59–61], artificial intelligence [62], big data analytics [63], clinical virtual reality [64], smartphones [65, 66], APPs [67], and so forth. The current paradigm shift in information and communication technologies (ICT) such as smart and ultra-compressed sensing, generative adversarial network (GAN), analog bio-inspired machine learning, trainable neuromorphic signal converting, and quantum computing, has been propelling the rapid pace of innovation in smart biofeedback. As a new technology regime of research and applications we only scratch the surface of smart biofeedback.

## 2. Perspectives and applications in smart biofeedback

This book addresses five important topics of the perspectives and applications in smart biofeedback—*Brain Networks*, *Neuromeditation*, *Psychophysiological Psychotherapy*, *Physiotherapy*, and *Privacy, Security, and Integrity of Data*, described as below.

### 2.1 Brain networks

Human brain is an information-sharing network which is connected by many spatially distributed, but functionally linked regions of the brain. Exploration of dynamic interactions of the large-scale brain network can construct the relationships among brain regions. It helps understand which and how brain regions actually cooperate during creative cognition and artistic performance. In past decades, neuroimaging studies have investigated the functional connectivity by measuring the level of co-activation of resting-state functional magnetic resonance imaging (fMRI) time series between brain regions [68]. Recent advancements in smart biofeedback technology have paved a promising avenue for real-time, *in vivo* analysis and modeling of nerve synapses in the brain. Electrical neuroimaging [69] that uses electroencephalography (EEG) biofeedback [70] as the neuroimaging technique has the advantages of online recording of the neuronal activity in real time. As a window into the brain, EEG has been used to localize the neural activity in the brain non-invasively. Biofeedback, also known as neurobiofeedback or neurofeedback (NFB) [71, 72], is a psychophysiological mechanism for training of self-regulation and has shown compelling findings about brain function and the neural correlates of behavior and cognition.

### 2.2 Neuromeditation

Over the last years, the soaring public interest in mindfulness meditation has raised scientific attention. For many people, mindfulness meditation brings a lot of benefits and is considered useful for training attention on cognition and emotion simultaneously. However, some of them are either difficult to maintain a disciplined practice regularly, or lack of methodologies or tools to develop their meditation practice more rapidly. Research studies have observed that many similarities of changes in EEG frequency bands trained in cognitive NFB therapeutic protocols are found in the mental activity involved in meditation practices [73]. As both meditation and NFB are techniques to train mental states, systems or applications based on NFB or real-time EEG biofeedback techniques have the potentials to help develop meditation. However, the reliability and accuracy of signal detection remain questionable in current neuromeditation applications. It is challenging to describe the complex brain activity during meditation by basic EEG analyses [74].

### 2.3 Psychophysiological psychotherapy

While psychophysiological and behavioral interventions manifest equivalent effectiveness for some kinds of psychological problems, most psychotherapeutic practices focus on the cognitive procedures. Even though psychophysiological approaches that use heart rate variability (HRV) biofeedback in muscle relaxation and breathing interventions have shown significant effects in clinical psychotherapies, psychophysiological methods are often downplayed [75, 76]. Psychotherapy is increasingly emphasizing clinical processes and mechanisms in developing smart biofeedback protocols in psychotherapeutic training programs to mitigate patient's physical and psychological disorders such as headache [77], depression [78], and anxiety [79]. As the engagement of an efficacious patient-therapist relationship is prominent in psychotherapy [80], development of innovative, artificial-intelligence-augmented tools and systems for interpersonal biofeedback [81] is required to optimize therapists' awareness of unconscious interpersonal regulation dynamics on a moment-to-moment basis.

### 2.4 Physiotherapy

Biofeedback has been an established, non-pharmacologic technique in physiotherapy and rehabilitation to mitigate migraine headache [82], relax pelvic floor muscles [83–86], help stoke patients regain movement in paralyzed muscles [87, 88], relieve chronic pains [89], and reduce Raynaud's phenomenon [90]. Recently, a combined approach, named Imagined Imitation [91], that integrates traditional physiotherapy with NBF appears to establish more therapeutic benefits. In conjunction with electromechanical physiotherapy, participants who receive NFB training sessions get better improved in motor control of their affected arms [92]. The mechanism of the effectiveness by combining NFB with physiotherapy to modulate brain activity is still yet to explore. However, such a combined approach opens a grand door to both researchers and practitioners for design of user interfaces and user experience scenarios which drive sound user engagement with proper match of the level of challenge, according to the participant's ability and conditions.

### 2.5 Privacy, security, and integrity of data

The prospectives and applications in smart biofeedback have to embrace the advanced ICT integration. In addition to diagnosis information, demographic, medical, and psychological data, as well as historical information such as duration of complaints and treatment history, should be well managed [93]. Simultaneous acquisition of data from multiple physiological sensors introduces new challenges to data management along the path of data life cycle of each data, from sensing, measuring, interpreting, transmitting, storing, analyzing, predicting, learning, to optimizing. Smart biofeedback belongs to highly interconnected systems and applications that demand end-to-end privacy, security, and integrity of data for proper operations [94]. Smart biofeedback systems and applications are all built on the assumption of reliable and secure communication but are vulnerably exposed to various treats and attacks. As more and more smart biofeedback systems and applications with IoT, cloud and edge computing implementation are emerging, new system architectures and approaches such as blockchains are to develop for massively distributed processing, to ensure the privacy, security, and integrity of data.

## 3. How the book is organized

This book consists of six chapters. The remaining of the book is into five sections to cope with the five specific topics of the perspectives and applications in smart biofeedback described above. All the chapters are from academic researchers and clinical practitioners on various areas of smart biofeedback. This introductory chapter, Chapter 1, describes the nature and purpose of this book and outlines the scope, logic and significance of the contents of this book.

Chapter 2 in the *Brain Networks* section overviews the recent advances in electrical neuroimaging, brain networks and neurofeedback protocols. It starts with a review of live Z-score NBF training which uses an EEG normative reference database to identify the possibly dysregulated brain regions from the probabilities estimated by auto and cross-spectrum of EEG. New advances in electrical neuro-imaging provide a 12,700-voxel resolution, three-dimensional EEG source location that uses swLORETA (weighted standardized low resolution electromagnetic tomography) and 19 channels for NBF. Such technologies enable the NFB approach to cerebellar and subcortical brain hubs like the thalamus, amygdala and habenula. Linking symptoms to dysregulated brain hubs and networks is crucial to help patients of NBF training. New development in cerebellar z-score NBF research has shown a bright future in NBF for brain networks. Potential applications of future swLORETA z-score NBF include helping people with cognition problems as well as balance problems, and Parkinsonism.

Chapter 3 in the *Neuromeditation* section introduces the science and practice of neuromeditation. Combining NFB and meditation, neuromeditation monitors brainwave activity to help meditators learn to quickly enter a desired state of consciousness and then maintain the state for a period of time for improvements in mental health. The history, examples, and research evidence on neuromeditation are reviewed. As researches on both NFB and meditation have found effective in the treatment of various mental health concerns, respectively, the efficacy of neuromeditation is reviewed with a case study about a middle-aged, Caucasian female with anxiety, eating disorder, and post traumatic stress disorder (PTSD) mental health history. Assessment results, including a quantitative EEG (qEEG) assessment comparing her baseline EEG activity to a clinical database, indicate that her concerns most closely match the Quiet Mind style of meditation. A mindfulness meditation protocol with eight neuromeditation sessions is identified as the best for her concerns and background. Most of the improvements are found directly related to the goals and concerns identified in the intake process.

Chapter 4 in the *Psychophysiological Psychotherapy* section surveys the most common eight modalities of biofeedback in clinical psychology and provides the perspectives of how biofeedback is applied to manage the psychophysiological conditions. It aims to elucidate the clinical settings in psychotherapy practices and point out the psychophysiological conditions for biofeedback training. It reviews the basic principles of psychophysiology, including the balance mechanism between the fight-or-flight (sympathetic) versus the rest-and-digest (parasympathetic) pathways in the autonomic nervous system (ANS) where biofeedback can regulate and provide balance between sympathetic and parasympathetic responses to mitigate psychopathological symptoms and to improve cognitive performance. It demonstrates several successful biofeedback training protocols in clinical psychology, including managing stress in anxious patients, managing hyperactivity of children with attention deficit hyperactivity disorder (ADHD), managing depression.

Chapter 5 in the *Physiotherapy* section deals with the problems of chronic pains. It surveys six NBF protocols to evaluate the efficacy of NBF training for treatment to chronic pains. At first, it introduces the mechanisms and neural pathways underlying pain perception, and describes the brain rhythms associated with chronic pains and the identification of neurophysiological correlates of chronic pains. The efficacy of the two key NBF modalities, EEG and fMRI, for chronic pains are discussed. A NBF protocol with EEG target brain rhythms is provided. It surveys the efficacy of various NFB training protocols applied in the management of chronic pains, including fibromyalgia, central neuropathic pains in paraplegic patients, traumatic brain injury, chemotherapy-induced peripheral neuropathy, primary headache, complex region pain syndrome type I, post-herpetic neuralgia, and chronic lower back pains. The feasibility and availability of home-based NFB therapy for chronic pains are reviewed. Finally, it discusses the adverse effects associated with NFB training and limitations, and provides future recommendations.

Chapter 6 in the *Privacy, Security, and Integrity of Data* section presents the design and applications of the Blockchain-based Medical Data Management System (BMDMS) for management of medical health records and biofeedback data. As an emerging and prospective technology in ICT, blockchain is featured by its privacy, security, and integrity of data so that blockchain-enabled information and data will remain safe and secure. BMDMS adopts the smart contract approach to reduce the need in trusted intermediators, fraud losses, and accidental or malicious exceptions, in order to protect the privacy of patients and medical practitioners. Patients can take biofeedback training at home or in any local clinics, hospitals, or large medical centers and their medical health records and collected biofeedback data can be shared among medical facilities within BMDMS in a secure and safe way. The proposed BMDMS framework utilizes big data, analytics, and edge/cloud computing technologies, to smartly retrieve the exact amount of data to optimize the computing and storage capabilities. BMDMS thus avoids the possible deluge of clinic or laboratory data, such as massive data collected from biofeedback sensors.

## Author details

Edward Da-Yin Liao
Straight and Up Intelligent Innovations Group Co., San Jose, CA, USA

*Address all correspondence to: eliao@miicg.com

## References

[1] Karlins, M. and Andrews, L.M., 1972. Biofeedback: Turning on the power of your mind. Lippincott Williams & Wilkins.

[2] Kater, D. and Spires, J., 1975. Biofeedback: The beat goes on. *The School Counselor*, *23*(1), pp.16-21.

[3] Tarler-Benlolo, L., 1978. The role of relaxation in biofeedback training: a critical review of the literature. *Psychological Bulletin*, *85*(4), p.727.

[4] Blanchard, E.B. and Young, L.D., 1974. Clinical applications of biofeedback training: A review of evidence. *Archives of General Psychiatry*, *30*(5), pp.573-589.

[5] Miller, N.E., 1978. Biofeedback and visceral learning. *Annual review of psychology*, *29*(1), pp.373-404.

[6] King, N.J. and Montgomery, R.B., 1980. Biofeedback-induced control of human peripheral temperature: A critical review of literature. *Psychological Bulletin*, *88*(3), p.738.

[7] Duckro, P.N. and Cantwell-Simmons, E., 1989. A review of studies evaluating biofeedback and relaxation training in the management of pediatric headache. *Headache: The Journal of Head and Face Pain*, *29*(7), pp.428-433.

[8] Futterman, A.D. and Shapiro, D., 1986. A review of biofeedback for mental disorders. *Psychiatric Services*, *37*(1), pp.27-33.

[9] Enck, P., 1993. Biofeedback training in disordered defecation. *Digestive diseases and sciences*, *38*(11), pp.1953-1960.

[10] Riegel, B., Warmoth, J.E., Middaugh, S.J., Kee, W.G., Nicholson, L.C., Melton, D.M., Parikh, D.K. and Rosenberg, J.C., 1995. Psychogenic cough treated with biofeedback and psychotherapy. A review and case report. *American journal of physical medicine & rehabilitation*, *74*(2), pp.155-158.

[11] Trudeau, D.L., 2000. The treatment of addictive disorders by brain wave biofeedback: A review and suggestions for future research. *Clinical Electroencephalography*, *31*(1), pp.13-22.

[12] Palsson, O.S., Heymen, S. and Whitehead, W.E., 2004. Biofeedback treatment for functional anorectal disorders: a comprehensive efficacy review. *Applied psychophysiology and biofeedback*, *29*(3), pp.153-174.

[13] Maryn, Y., De Bodt, M. and Van Cauwenberge, P., 2006. Effects of biofeedback in phonatory disorders and phonatory performance: a systematic literature review. *Applied psychophysiology and biofeedback*, *31*(1), pp.65-83.

[14] Medlicott, M.S. and Harris, S.R., 2006. A systematic review of the effectiveness of exercise, manual therapy, electrotherapy, relaxation training, and biofeedback in the management of temporomandibular disorder. *Physical therapy*, *86*(7), pp.955-973.

[15] Nestoriuc, Y., Martin, A., Rief, W. and Andrasik, F., 2008. Biofeedback treatment for headache disorders: a comprehensive efficacy review. *Applied psychophysiology and biofeedback*, *33*(3), pp.125-140.

[16] Sokhadze, T.M., Cannon, R.L. and Trudeau, D.L., 2008. EEG biofeedback as a treatment for substance use disorders: review, rating of efficacy, and recommendations for further research. *Applied psychophysiology and biofeedback*, *33*(1), pp.1-28.

[17] Schoenberg, P.L. and David, A.S., 2014. Biofeedback for psychiatric disorders: a systematic review. *Applied psychophysiology and biofeedback*, *39*(2), pp.109-135.

[18] Fazeli, M.S., Lin, Y., Nikoo, N., Jaggumantri, S., Collet, J.P. and Afshar, K., 2015. Biofeedback for nonneuropathic daytime voiding disorders in children: a systematic review and meta-analysis of randomized controlled trials. *The Journal of urology*, *193*(1), pp.274-280.

[19] Tremback-Ball, A., Gherghel, E., Hegge, A., Kindig, K., Marsico, H. and Scanlon, R., 2018. The effectiveness of biofeedback therapy in managing Bladder Bowel Dysfunction in children: A systematic review. *Journal of pediatric rehabilitation medicine*, *11*(3), pp.161-173.

[20] Jokubauskas, L. and Baltrušaitytė, A., 2018. Efficacy of biofeedback therapy on sleep bruxism: A systematic review and meta-analysis. *Journal of oral rehabilitation*, *45*(6), pp.485-495.

[21] Sugden, E., Lloyd, S., Lam, J. and Cleland, J., 2019. Systematic review of ultrasound visual biofeedback in intervention for speech sound disorders. *International journal of language & communication disorders*, *54*(5), pp.705-728.

[22] Basmajian, J.V., 1981. Biofeedback in rehabilitation: a review of principles and practices. *Archives of physical medicine and rehabilitation*, *62*(10), pp.469-475.

[23] Cozean, C.D., Pease, W.S. and Hubbell, S.L., 1988. Biofeedback and functional electric stimulation in stroke rehabilitation. *Archives of physical medicine and rehabilitation*, *69*(6), pp.401-405.

[24] Glanz, M., Klawansky, S. and Chalmers, T., 1997. Biofeedback therapy in stroke rehabilitation: a review. *Journal of the Royal Society of Medicine*, *90*(1), pp.33-39.

[25] Huang, H., Wolf, S.L. and He, J., 2006. Recent developments in biofeedback for neuromotor rehabilitation. *Journal of neuroengineering and rehabilitation*, *3*(1), p.11.

[26] Giggins, O.M., Persson, U.M. and Caulfield, B., 2013. Biofeedback in rehabilitation. *Journal of neuroengineering and rehabilitation*, *10*(1), p.60.

[27] Prinsloo, G.E., Rauch, H.L. and Derman, W.E., 2014. A brief review and clinical application of heart rate variability biofeedback in sports, exercise, and rehabilitation medicine. *The Physician and sportsmedicine*, *42*(2), pp.88-99.

[28] Richards, R., van den Noort, J.C., Dekker, J. and Harlaar, J., 2017. Gait retraining with real-time biofeedback to reduce knee adduction moment: systematic review of effects and methods used. *Archives of physical medicine and rehabilitation*, *98*(1), pp.137-150.

[29] Andrasik, F., 2010. Biofeedback in headache: an overview of approaches and evidence. *Cleve Clin J Med*, *77*(Suppl 3), pp.S72-S76.

[30] Klich, U., 2015. The integration of mindfulness-based biofeedback and compassion in the healthcare setting. *Biofeedback*, *43*(3), pp.111-116.

[31] Badawi, H.F. and El Saddik, A., 2020. Biofeedback in healthcare: State of the art and meta review. In *Connected Health in Smart Cities* (pp. 113-142). Springer, Cham.

[32] Zijlstra, A., Mancini, M., Chiari, L. and Zijlstra, W., 2010. Biofeedback for training balance and mobility tasks in older populations: a systematic

review. *Journal of neuroengineering and rehabilitation*, *7*(1), p.58.

[33] Ma, C.Z.H., Wong, D.W.C., Lam, W.K., Wan, A.H.P. and Lee, W.C.C., 2016. Balance improvement effects of biofeedback systems with state-of-the-art wearable sensors: A systematic review. *Sensors*, *16*(4), p.434.

[34] Liao, D.Y., 2016, June. Design of a secure, biofeedback, head-and-neck posture correction system. In 2016 IEEE First International Conference on Connected Health: Applications, Systems and Engineering Technologies (CHASE) (pp. 119-124). IEEE.

[35] Cerqueira, S.M., Da Silva, A.F. and Santos, C.P., 2020. Smart Vest for Real-Time Postural Biofeedback and Ergonomic Risk Assessment. *IEEE Access*, *8*, pp.107583-107592.

[36] Yu, B., Hu, J., Funk, M., Liang, R.H., Xue, M. and Feijs, L., 2018. RESonance: Lightweight, room-scale audio-visual biofeedback for immersive relaxation training. *IEEE Access*, *6*, pp.38336-38347.

[37] Umek, A., Zhang, Y., Tomažič, S. and Kos, A., 2017. Suitability of strain gage sensors for integration into smart sport equipment: A golf club example. *Sensors*, *17*(4), p.916.

[38] Raymond, J., Sajid, I., Parkinson, L.A. and Gruzelier, J.H., 2005. Biofeedback and dance performance: A preliminary investigation. *Applied psychophysiology and biofeedback*, *30*(1), pp.65-73.

[39] Liao, D.Y., 2018. Collaborative, Social-Networked Posture Training with Posturing Monitoring and Biofeedback. In *Biofeedback*. IntechOpen.

[40] Ciolacu, M.I., Binder, L. and Popp, H., 2019, October. Enabling IoT in Education 4.0 with BioSensors from Wearables and Artificial Intelligence. In *2019 IEEE 25th International Symposium for Design and Technology in Electronic Packaging (SIITME)* (pp. 17-24). IEEE.

[41] Weatherall, M., 1999. Biofeedback or pelvic floor muscle exercises for female genuine stress incontinence: a meta-analysis of trials identified in a systematic review. *BJU international*, *83*, pp.1015-1016.

[42] Peake, J.M., Kerr, G. and Sullivan, J.P., 2018. A critical review of consumer wearables, mobile applications, and equipment for providing biofeedback, monitoring stress, and sleep in physically active populations. *Frontiers in physiology*, *9*, p.743.

[43] Kennedy, L. and Parker, S.H., 2019. Biofeedback as a stress management tool: a systematic review. *Cognition, Technology & Work*, *21*(2), pp.161-190.

[44] De Witte, N.A., Buyck, I. and Van Daele, T., 2019. Combining biofeedback with stress management interventions: A systematic review of physiological and psychological effects. *Applied Psychophysiology and Biofeedback*, *44*(2), pp.71-82.

[45] Horta, E.T., Lopes, I.C., Rodrigues, J.J. and Misra, S., 2013, October. Real time falls prevention and detection with biofeedback monitoring solution for mobile environments. In *2013 IEEE 15th International Conference on e-Health Networking, Applications and Services (Healthcom 2013)* (pp. 594-600). IEEE.

[46] Nacke, L.E., Kalyn, M., Lough, C. and Mandryk, R.L., 2011, May. Biofeedback game design: using direct and indirect physiological control to enhance game interaction. In *Proceedings of the SIGCHI conference on human factors in computing systems* (pp. 103-112).

[47] Mavroidis, C., Nikitczuk, J., Weinberg, B., Arango, R., Danaher, G., Jensen, K., Leahey, M., Pavone, R.,

Pelletier, P., Provo, A. and Prugnarola, J., 2005, January. Smart portable rehabilitation devices. In *International Design Engineering Technical Conferences and Computers and Information in Engineering Conference* (Vol. 47446, pp. 501-510).

[48] Senanayake, S.A. and Naim, A.G., 2019. Smart sensing and biofeedback for vertical jump in sports. In *Modern Sensing Technologies* (pp. 63-81). Springer, Cham.

[49] Anandh, S., Vasuki, D.R., Al Baradie, D. and Saleem, R., 2020. Smart Biofeedback Expectorant System for Improving the Lung Capacities. *International Journal of Advanced Research in Engineering and Technology (IJARET)*, *11*(3), pp.240-248.

[50] Kos, A. and Umek, A., 2018. Smart sport equipment: SmartSki prototype for biofeedback applications in skiing. *Personal and Ubiquitous Computing*, *22*(3), pp.535-544.

[51] Shen, T.W., Hsiao, T., Liu, Y.T. and He, T.Y., 2008, November. An ear-lead ECG based smart sensor system with voice biofeedback for daily activity monitoring. In *Tencon 2008-2008 IEEE Region 10 Conference* (pp. 1-6). IEEE.

[52] Meriggi, P., Mandalà, M., Brazzoli, E., Piacente, T., Mazzola, M. and Olivieri, I., 2018, July. Smart objects and biofeedback for a pediatric rehabilitation 2.0. In *Italian Forum of Ambient Assisted Living* (pp. 105-119). Springer, Cham.

[53] Johansson, A., Shen, W. and Xu, Y., 2011, September. An ANT based wireless body sensor biofeedback network for medical e-health care. In *2011 7th International Conference on Wireless Communications, Networking and Mobile Computing* (pp. 1-5). IEEE.

[54] Kos, A., Milutinović, V. and Umek, A., 2019. Challenges in wireless communication for connected sensors and wearable devices used in sport biofeedback applications. *Future generation computer systems*, *92*, pp.582-592.

[55] De Venuto, D., Annese, V.F. and Sangiovanni-Vincentelli, A.L., 2016, May. The ultimate IoT application: A cyber-physical system for ambient assisted living. In *2016 IEEE International Symposium on Circuits and Systems (ISCAS)* (pp. 2042-2045). IEEE.

[56] Rastogi, R., Chaturvedi, D.K. and Gupta, M., 2020. Tension Type Headache: IOT and FOG Applications in Healthcare Using Different Biofeedback. In *Handbook of Research on Advancements of Artificial Intelligence in Healthcare Engineering* (pp. 318-358). IGI Global.

[57] Umek, A., Tomažič, S. and Kos, A., 2015. Wearable training system with real-time biofeedback and gesture user interface. *Personal and Ubiquitous Computing*, *19*(7), pp.989-998.

[58] Marta, G., Simona, F., Andrea, C., Dario, B., Stefano, S., Federico, V., Marco, B., Francesco, B., Stefano, M. and Alessandra, P., 2019. Wearable biofeedback suit to promote and monitor aquatic exercises: a feasibility study. *IEEE Transactions on Instrumentation and Measurement*, *69*(4), pp.1219-1231.

[59] Rodrigues, J.J., Pereira, O.R. and Neves, P.A., 2011. Biofeedback data visualization for body sensor networks. *Journal of Network and Computer Applications*, *34*(1), pp.151-158.

[60] Caetano, I., Alves, J., Gonçalves, J., Martins, M. and Santos, C.P., 2016, May. Development of a biofeedback approach using body tracking with Active Depth sensor in ASBGo smart walker. In *2016 International Conference on Autonomous Robot Systems and Competitions (ICARSC)* (pp. 241-246). IEEE.

[61] Wang, Y., Wan, B., Li, H. and Shan, G., 2016. A wireless sensor system for a biofeedback training of hammer throwers. *SpringerPlus*, *5*(1), pp.1-14.

[62] Mongan, W.M., 2018. Predictive Analytics on Real-Time Biofeedback for Actionable Classification of Activity State. Drexel University.

[63] Janatova, M., Uller, M., Stepankova, O., Brezany, P. and Lenart, M., 2019. A Novel Big Data-Enabled Approach, Individualizing and Optimizing Brain Disorder Rehabilitation. In *Big Data for the Greater Good* (pp. 101-127). Springer, Cham.

[64] Repetto, C., Gorini, A., Vigna, C., Algeri, D., Pallavicini, F. and Riva, G., 2009. The use of biofeedback in clinical virtual reality: the INTREPID project. *JoVE (Journal of Visualized Experiments)*, (33), p.e1554.

[65] Franco, C., Fleury, A., Guméry, P.Y., Diot, B., Demongeot, J. and Vuillerme, N., 2012. iBalance-ABF: a smartphone-based audio-biofeedback balance system. *IEEE transactions on biomedical engineering*, *60*(1), pp.211-215.

[66] Kos, A., Tomažič, S. and Umek, A., 2016. Suitability of smartphone inertial sensors for real-time biofeedback applications. *Sensors*, *16*(3), p.301.

[67] Rastogi, R., Chaturvedi, D.K., Gupta, M. and Singhal, P., 2020. Intelligent Mental Health Analyzer by Biofeedback: App and Analysis. In *Handbook of Research on Optimizing Healthcare Management Techniques* (pp. 127-153). IGI Global.

[68] Van Den Heuvel, M.P. and Pol, H.E.H., 2010. Exploring the brain network: a review on resting-state fMRI functional connectivity. *European neuropsychopharmacology*, *20*(8), pp.519-534.

[69] Michel, C.M., Koenig, T., Brandeis, D., Wackermann, J. and Gianotti, L.R. eds., 2009. *Electrical neuroimaging*. Cambridge University Press.

[70] Enriquez-Geppert, S., Huster, R.J. and Herrmann, C.S., 2017. EEG-neurofeedback as a tool to modulate cognition and behavior: a review tutorial. *Frontiers in human neuroscience*, *11*, p.51.

[71] Hammond, D.C., 2007. What is neurofeedback?. *Journal of neurotherapy*, *10*(4), pp.25-36.

[72] Sitaram, R., Ros, T., Stoeckel, L., Haller, S., Scharnowski, F., Lewis-Peacock, J., Weiskopf, N., Blefari, M.L., Rana, M., Oblak, E. and Birbaumer, N., 2017. Closed-loop brain training: the science of neurofeedback. *Nature Reviews Neuroscience*, *18*(2), pp.86-100.

[73] Cahn, B.R., Delorme, A. and Polich, J., 2013. Event-related delta, theta, alpha and gamma correlates to auditory oddball processing during Vipassana meditation. *Social cognitive and affective neuroscience*, *8*(1), pp.100-111.

[74] Travis, F. and Shear, J., 2010. Focused attention, open monitoring and automatic self-transcending: categories to organize meditations from Vedic, Buddhist and Chinese traditions. *Consciousness and cognition*, *19*(4), pp.1110-1118.

[75] Lehrer, P., 2017. Biofeedback: An important but often-ignored ingredient in psychotherapy. *Policy Insights from the Behavioral and Brain Sciences*, *4*(1), pp.57-63.

[76] Morgan, S.J. and Mora, J.A.M., 2017. Effect of heart rate variability biofeedback on sport performance, a systematic review. *Applied psychophysiology and biofeedback*, *42*(3), pp.235-245.

[77] Jessup, B.A., Neufeld, R.W. and Merskey, H., 1979. Biofeedback therapy for headache and other pain: An evaluative review. *Pain*, *7*(3), pp.225-270.

[78] Caldwell, Y.T. and Steffen, P.R., 2018. Adding HRV biofeedback to psychotherapy increases heart rate variability and improves the treatment of major depressive disorder. *International Journal of Psychophysiology*, *131*, pp.96-101.

[79] Moore, N.C., 2000. A review of EEG biofeedback treatment of anxiety disorders. *Clinical electroencephalography*, *31*(1), pp.1-6.

[80] Kleinbub, J.R., Mannarini, S. and Palmieri, A., 2020. Interpersonal Biofeedback in Psychodynamic Psychotherapy. *Frontiers in Psychology*, *11*.

[81] Kassel, S.C. and LeMay, J., 2012. Interpersonal biofeedback. *The Therapist*, pp.68-70.

[82] Nestoriuc, Y. and Martin, A., 2007. Efficacy of biofeedback for migraine: a meta-analysis. *Pain*, *128*(1-2), pp.111-127.

[83] Bassotti, G., Chistolini, F., Sietchiping-Nzepa, F., De Roberto, G., Morelli, A. and Chiarioni, G., 2004. Biofeedback for pelvic floor dysfunction in constipation. *Bmj*, *328*(7436), pp.393-396.

[84] Glazer, H.I. and Laine, C.D., 2006. Pelvic floor muscle biofeedback in the treatment of urinary incontinence: A literature review. *Applied Psychophysiology and Biofeedback*, *31*(3), pp.187-201.

[85] Fitz, F.F., Resende, A.P.M., Stüpp, L., Sartori, M.G.F., Girão, M.J.B.C. and Castro, R.A., 2012. Biofeedback for the treatment of female pelvic floor muscle dysfunction: a systematic review and meta-analysis. *International urogynecology journal*, *23*(11), pp.1495-1516.

[86] Herderschee, R., Hay-Smith, E.J., Herbison, G.P., Roovers, J.P. and Heineman, M.J., 2013. Feedback or biofeedback to augment pelvic floor muscle training for urinary incontinence in women: shortened version of a Cochrane systematic review. *Neurourology and urodynamics*, *32*(4), pp.325-329.

[87] Wolf, S.L., 1983. Electromyographic biofeedback applications to stroke patients: a critical review. *Physical therapy*, *63*(9), pp.1448-1459.

[88] Nelson, L.A., 2007. The role of biofeedback in stroke rehabilitation: past and future directions. *Topics in stroke rehabilitation*, *14*(4), pp.59-66.

[89] Jensen, M.P., Gertz, K.J., Kupper, A.E., Braden, A.L., Howe, J.D., Hakimian, S. and Sherlin, L.H., 2013. Steps toward developing an EEG biofeedback treatment for chronic pain. *Applied psychophysiology and biofeedback*, *38*(2), pp.101-108.

[90] Frank, D.L., Khorshid, L., Kiffer, J.F., Moravec, C.S. and McKee, M.G., 2010. Biofeedback in medicine: who, when, why and how?. *Mental health in family medicine*, *7*(2), p.85.

[91] Friesen, C.L., Bardouille, T., Neyedli, H.F. and Boe, S.G., 2017. Combined action observation and motor imagery neurofeedback for modulation of brain activity. *Frontiers in human neuroscience*, *10*, p.692.

[92] Ramos-Murguialday, A., Broetz, D., Rea, M., Läer, L., Yilmaz, Ö., Brasil, F.L., Liberati, G., Curado, M.R., Garcia-Cossio, E., Vyziotis, A. and Cho, W., 2013. Brain–machine interface in chronic stroke rehabilitation:

a controlled study. *Annals of neurology*, *74*(1), pp.100-108.

[93] Crider, A., Glaros, A.G. and Gevirtz, R.N., 2005. Efficacy of biofeedback-based treatments for temporomandibular disorders. *Applied psychophysiology and biofeedback*, *30*(4), pp.333-345.

[94] Pearson, S., 2013. Privacy, security and trust in cloud computing. In *Privacy and security for cloud computing* (pp. 3-42). Springer, London.

Section 2

# Brain Networks

Chapter 2

# Advances in Electrical Neuroimaging, Brain Networks and Neurofeedback Protocols

*Robert W. Thatcher, Carl J. Biver, Ernesto Palermero Soler, Joel Lubar and J. Lucas Koberda*

## Abstract

Human EEG biofeedback (neurofeedback) started in the 1940s using 1 EEG recording channel, then to 4 channels in the 1990s. New advancements in electrical neuroimaging expanded EEG biofeedback to 19 channels using Low Resolution Electromagnetic Tomography (LORETA) three-dimensional current sources of the EEG. In 2004–2006 the concept of a "real-time" comparison of the EEG to a healthy reference database was developed and tested using surface EEG z-score neurofeedback based on a statistical bell curve called "real-time" z-scores. The "real-time" or "live" normative reference database comparison was developed to help reduce the uncertainty of what threshold to select to activate a feedback signal and to unify all EEG measures to a single value, i.e., the distance from the mean of an age matched reference sample. In 2009 LORETA z-score neurofeedback further increased the specificity by targeting brain network hubs referred to as Brodmann areas. A symptom check list program to help link symptoms to dysregulation of brain networks based on fMRI and PET and neurology was created in 2009. The symptom checklist and NIH based networks linking symptoms to brain networks grew out of the human brain mapping program starting in 1990 which is continuing today. A goal is to increase specificity of EEG biofeedback by targeting brain network hubs and connections between hubs likely linked to the patient's symptoms. New advancements in electrical neuroimaging introduced in 2017 provide increased resolution of three-dimensional source localization with 12,700 voxels using swLORETA with the capacity to conduct cerebellar neurofeedback and neurofeedback of subcortical brain hubs such as the thalamus, amygdala and habenula. Future applications of swLORETA z-score neurofeedback represents another example of the transfer of knowledge gained by the human brain mapping initiatives to further aid in helping people with cognition problems as well as balance problems and parkinsonism. A brief review of the past, present and future predictions of z-score neurofeedback are discussed with special emphasis on new developments that point toward a bright and enlightened future in the field of EEG biofeedback.

**Keywords:** QEEG, neurofeedback, cerebellar EEG, Z scores, swLORETA

## 1. History: raw scores to Z-scores

As previously published [1], normative reference databases serve a vital and important function in modern clinical science and patient evaluation, including quantitative EEG (QEEG) (see review by Thatcher and Lubar [2]. Clinical normative databases aid in the evaluation of a wide range of disorders by using statistics to estimate the distance from the mean of an age matched normal reference. For example, blood constituent normative databases, MRI, fMRI and Positron emission tomography (PET), ocular and retinal normative databases, blood pressure normative databases, nerve conduction velocity normative databases, postural databases, bone density normative databases, ultra sound normative databases and motor development normative databases, to name a few. A comprehensive survey of existing clinical normative databases can be obtained by searching the National Library of Medicine's database using the search term "Normative Databases" at http://www.ncbi.nlm.nih.gov/sites/entrez.

In 1998 the fundamental design concept of real-time Z score biofeedback was to use a EEG normative database from birth to old age from a reference group of healthy individuals like a real-time blood test comparison to a blood constituent normative database but instead it is a EEG normative database [3–6]. The central idea was a real-time z-score using the standard bell curve by which probabilities for an individual can be estimated using the auto and cross-spectrum of the electroencephalogram (EEG) in order to identify brain regions that are dysregulated and depart from expected values. While one- to four-channel z-score biofeedback is valuable, the linkage of symptoms and complaints to functional network hubs in the brain is best achieved by the use of 19 channels of EEG to compare a patient's EEG to the fMRI and PET human brain mapping studies linked to brain networks and using an age matched normative database so that current source localization in Brodmann areas (network hubs) and connections between network hubs can be computed. Once the linkage is made of symptoms to the weak hubs and connections likely linked to symptoms, then an individualized z-score biofeedback protocol can be devised. However, in order to compute a z-score to make a linkage to symptoms then an accurate statistical inference must be made using the Gaussian distribution (i.e., bell curve).

Clinically applied normative databases share a common set of statistical and scientific standards that have evolved over the years. The standards include peer-reviewed publications, disclosure of the inclusion/exclusion criteria, tests of statistical validity, tests of reliability, cross-validation tests, adequate sample sizes for different age groups, etc. Normative databases are distinct from nonclinical control groups in their scope and their sampling restriction to clinically normal or otherwise healthy individuals for the purpose of comparison. Another distinguishing characteristic of normative databases is the ability to compare a single individual to a population of "normal" individuals in order to identify the measures that are distant from normal and the magnitude of deviation. Normative databases themselves do not diagnose a patient's clinical problem. Rather, a trained professional first evaluates the patient's clinical history and clinical symptoms and complaints and then uses the results of normative database comparisons to aid in the development of an accurate clinical diagnosis.

The real-time EEG z-score is directly related to the sample size for a given age group and the variance of the reference normal population distribution at each age. However, in order to achieve a representative Gaussian distribution, it is necessary to include two major categories of statistical variance: 1) the moment-to-moment variance or within-session variance, and 2) between subject variance across an age group. In the case of the fast Fourier transform (FFT) there is a single integral of the

*DOI: http://dx.doi.org/10.5772/intechopen.94326*

power spectrum for each subject and each frequency, and therefore, there is only between-subject variance in normative databases that use non-instantaneous analyses such as the FFT. The application of a normative database by the use of the FFT is recommended to start with symptoms and then to reject or confirm hypotheses about brain regions and networks by assessing the EEG, and thereby to then create a neurofeedback protocol linked to the patient's symptoms. Unlike the FFT, the Joint-Time-Frequency-Analysis (JTFA) z-score is computed in microseconds limited by the sample rate of the EEG amplifier; therefore, they are essentially instantaneous z-scores. It is necessary under the principals of operant conditioning that contiguity not be too fast because the activation of neuromodulators like dopamine are relatively slow and long-lasting [7, 8]. Therefore, 250 msec to about 1 second are commonly used intervals between a brain event that meets threshold and the delivery of a reinforcing signal for both raw score and z-score EEG biofeedback.

As illustrated in **Figure 1**, another design concept is simplification and standardization of EEG biofeedback by the application of basic science. Simplification is achieved by the use of a single metric, namely, the metric of the z-score for widely diverse measures such as power, amplitude asymmetry, power ratios, coherence, phase delays, phase-slope-index, phase reset, etc. A virtue of a z-score is metric independence and therefore there is no need to argue about absolute thresholds e.g., is it 30 μV or maybe 5 μV or maybe 15 μV, or should coherence be 0.6 or perhaps 0.9, or phase difference 25° or 62° or 110°, etc.? In addition to removing the guesswork, there is also no need to inhibit theta and reinforce beta, since both occur at the same time. That is, reinforcing toward $z = 0$ is a common goal whether dysregulation is a negative or a positive outlier because they are treated the same; i.e., the event is not reinforced if deviant from normal or distant from $z = 0$. Artifact rejection is another automatic feature of z-score neurofeedback. For example,

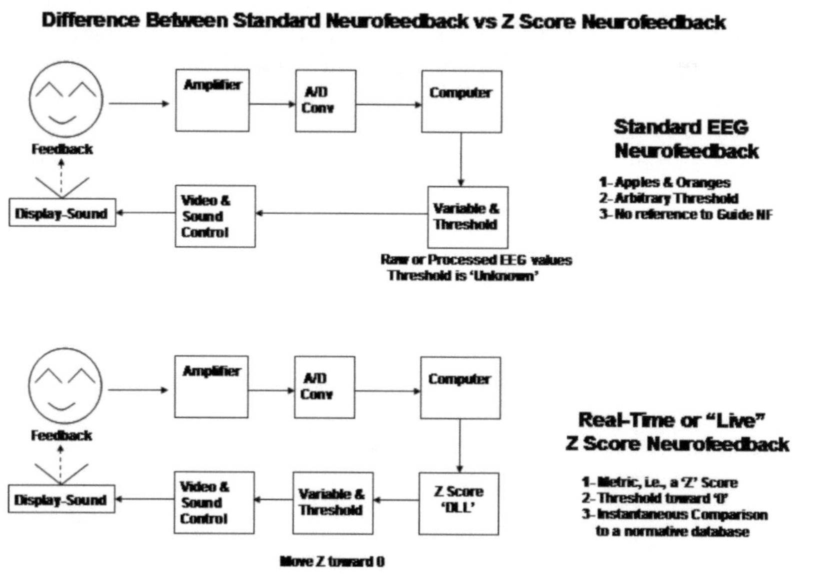


**Figure 1.**
*Top row is conventional or standard EEG biofeedback in which different units of measurement are used in an EEG analysis (e.g., μV for amplitude, theta/beta ratios, relative power 0 to 100%, coherence 0 to 1, phase in degrees or radians, etc.) and the clinician must guess at a threshold for a particular electrode location and frequency and age for when to reinforce or inhibit a give measure. The bottom row is z-score biofeedback, in which different metrics are represented by a single and common metric, i.e., the metric of a z-score, and the guesswork is removed because all measures are reinforced to move z-scores toward z = 0, which is the approximate center of an average healthy brain state based on a reference age-matched normative database in real time. Reprinted with permission from [9]). (Copyright 2012 Anipublishing, Inc.).*

artifact is usually 5 to 20 standard deviations from the non-artifact reference normative means and standard deviations, and if the reinforcement range is + and – 2 standard deviations, then artifact will not be reinforced, in contrast to raw score neurofeedback where movement and EMG artifact, etc. may be reinforced. Standardization is also achieved by EEG amplifier-matching of the frequency response of the normative database amplifiers to the frequency characteristics of the EEG amplifiers used to acquire a comparison to a subject's EEG time series. Without amplifier matching then deviation from normal may be because of the amplifier and not the patient's brain. This is one of the reasons that an amplifier-matched EEG normative database met FDA standards [9, 10].

### 1.1 Advances in EEG source localization

EEG source localization was developed in the 1980s and supported by the Human Brain Mapping program at the National Institutes of Health starting in 1990 and continuing today. Numerous cross-validations and tests of localization accuracy have been conducted and are reviewed in Thatcher [9, 10]. LORETA using 2394 MRI voxels was developed by Pascual-Marqui and colleagues in 1994 [11]. An improved version based on standardization of the source space and using 6200 MRI voxels was developed in 2003 called sLORETA. A limitation of LORETA and sLORETA is the reliance on a spherical head model because the brain is shaped like a loaf of bread, elongated and flat on the bottom, and it is not shaped like a sphere. In addition, the volume in the interior of the brain is not homogeneous, which results in reduced localization accuracy. In 2007, Ernesto Palmero Soler [12] developed an improved inverse solution by mathematically transforming the heterogeneous volume conductor into a homogeneous volume conductor and also by not using a spherical head model. Instead, Soler et al. [12] used a realistic head model using the more precise boundary element method (BEM) as well as 12,700 MRI voxels. This method is referred to as swLORETA or weighted sLORETA. The BEM plus the use of a homogeneous volume conduction results in improved source localization accuracy of deeper sources such as from the cerebellum and subthalamus and thalamus, etc. [13, 14].

**Figure 2** illustrates the SVD matrix operation to transform the heterogenous electrical lead field into a homogenous lead field. **Figure 3** shows the results of

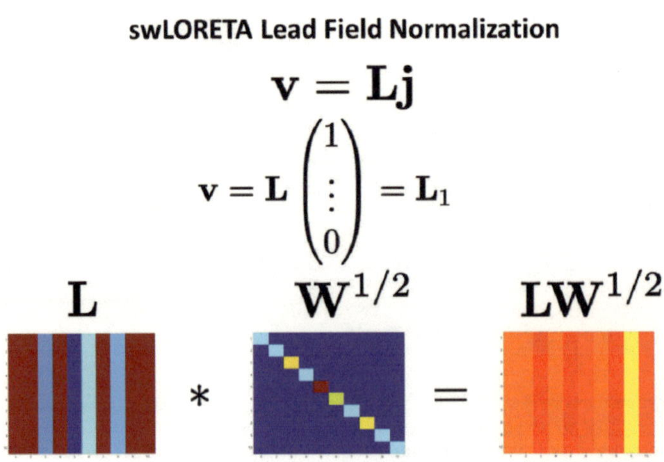


**Figure 2.**
*Top is the equation for the inverse solution, v = voltage, L = lead field and j = source currents. The SVD weighting matrix transforms L into $L_1$ (middle). Bottom row illustrates the transform of the heterogeneous lead field L by SVD to produce the homogeneous lead field $LW^{1/2}$.*

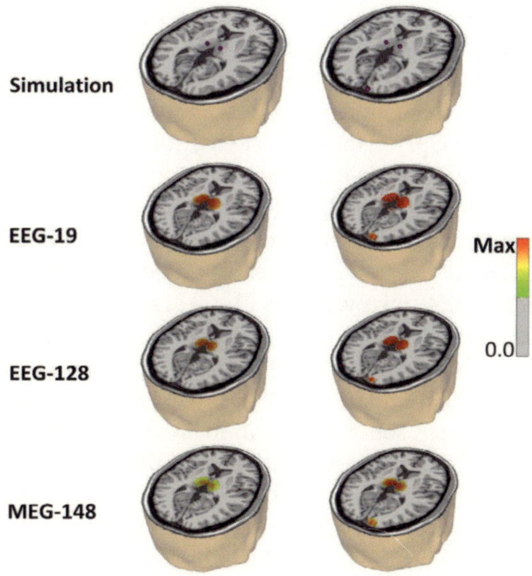


**Figure 3.**
*Simulations of the cerebral activity by deep sources are simulated using a linear combination of sine functions with frequency components evenly spaced in the alpha band (8–12 Hz). The amplitude of oscillation was the same for all the frequencies and it was set to 1.0. The 19 channels use the 10–20 positions electrodes system, the 128 use the 10–10 system and the MEG 148 follows the magnetometer configuration of the 4D neuroimaging MAGNES 2500 WH system. In this system, 148 magnetometers are arranged in a uniformly distributed array with a mean inter-channel spacing of 2.9 cm. .Left are two thalamic sources located at Talairach coordinates [−10–20 8] and [10–20 8]. Right is the same thalamic sources plus a right hemisphere occipital source located at [17–100 5]. The error for the thalamic sources in both configurations are EEG −19 = 20 mm; EEG −128 = 18 mm; MEG −148 = 14 mm, while for the occipital source the error range from EEG – 19 = 7 mm; EEG −128 = 7 mm, MEG = 5 mm.*

simulations that compared localization accuracy with different numbers of sensors for EEG and MEG source localization.

As mentioned previously swLORETA uses a singular value decomposition lead field weighting that compensates for varying sensitivity of the sensors to current sources at different depths [12–14]. Also a realistic boundary element model (BEM) was used for solving the forward problem [15]. The solution was computed using 12,300 voxels (5.00-mm grid spacing) and it was restricted to the gray matter of cerebrum and cerebellum and cerebellar relay nuclei, i.e., red nu., sub-thalamus, thalamus. The locations were based on the probabilistic brain tissue maps available from the Montreal Neurological Institute [16, 17]. Talairach coordinates were obtained for every voxel by placing the corresponding Talairach markers onto the anatomical template [18]. The final coordinates of the maxima values (x,y,z, Talairach coordinates) provided for labeling the corresponding brain areas were based on the Talairach atlas. For the definition of cerebellar regions, we used the nomenclature of the MRI Atlas of the Human Cerebellum of Schmahmann [19]. In order to reduce the number of variables, adjacent frequency 0.5 Hz bins were averaged to produce a 1 Hz bin from 1 Hz to 40 Hz for each of the 12,300 gray matter voxels.

### 1.2 Accuracy of 19 channel EEG inverse solution

The accuracy of the inverse solution as a function of the density of EEG scalp electrodes has been discussed extensively since the 1990s with the beginning of the NIH human Brain Mapping Project [20]. Low Resolution Electromagnetic Tomography (LORETA) was developed in 1994 by Pascual et al. (1994) using 19 channel EEG recordings and since this time hundreds of 19 channel LORETA studies have been published. Pascual-Marqui [21] compared five state-of-the-art parametric algorithms which are the minimum norm (MN), weighted minimum norm (WMN), Low resolution electromagnetic tomography (LORETA), Backus-Gilbert and Weighted Resolution Optimization (WROP). Using a three-layer spherical head model with 818 grid points (intervoxel distance of 0.133) and 148 electrodes, the results showed that on average only LORETA has an acceptable localization error of 1 grid unit when simulating a scenario with a single source. When comparing MN solutions and LORETA solutions with different Lp norms, Yao and Dewald [22] have also found out that LORETA with the L1 norm gives the best overall estimation. Grech et al. [23] conducted extensive cross-validation and accuracy tests of LORETA, sLORETA, MN, WMN and SLF (Shrinking LORETA FOCUSS) using both regularization and no-regularization and two different measures of error.

Songa et al. [24] compared source localization between 19, 32, 64, 128 and 256 channel EEG recordings. Standardized LORETA (sLORETA) was significantly more accurate than the minimum norm (MN) for all comparisons and there was a modest reduction in localization error using sLORETA but no significant differences in spatial spread nor amplitude estimates [24]. A limitation of the Songa et al. [24] study was not using the Boundary Element Method (BEM) to compute a realistic head model f24or sLORETA. For example, Songa et al. ([24], p. 20) stated: "With sLORETA standardization, if there is an exact match between the head parameters (geometry & conductivity) that generate the head surface potentials and the head model that is employed for the forward model, sampling density and coverage does not matter, and perfect (with no noise) source reconstruction is guaranteed. With increasing accuracy of head conductivity models that match the individual subject, standardization methods (like sLORETA) may become defensible." As explained in Section 2.6, the current study not only used the BEM but, more importantly, used the method of single-value-decomposition (SVD) to eliminate the heterogeneity of the source space and thereby better approximate the zero error properties of sLORETA that Songa et al. [24] discuss.

The improved localization accuracy of cerebellar sources in the present study and by Cebolla et al. [13, 14] when using swLORETA is due to both the use of BEM and the use of single-value-decompensation (SVD) to transform the heterogeneous electrical lead field into a homogeneous lead field similar to the magnetic electroencephalography (MEG) lead field as shown in the bottom row of **Figure 3**. **Figure 3** shows the comparison between 19 channels and 128 channel EEG in a dipole simulation test [25]. The left column is with two thalamic sources and the right column includes one additional source in the right occipital cortex.

The EEG sources were simulated using a linear combination of sine functions with frequency components evenly spaced in the alpha band (8–12 Hz). The amplitude of oscillation was the same for all the frequencies and it was set to 1.0. In this study we used two source configurations (see **Figure 4**). The first configuration consists of two thalamic sources located at Talairach coordinates [−10–20 8] and [10–20 8]. The second configuration consists of the same thalamic sources as in the left configuration plus an occipital source located at Talairach coordinates [17–100 5]. The error for the thalamic sources in both configurations are EEG −19 = 20 mm; EEG −128 = 18 mm; MEG −148 = 14 mm, while for the occipital source the error ranged from EEG – 19 = 7 mm; EEG −128 = 7 mm; MEG = 5 mm. Therefore, the simulation showed similar localization accuracy between 19 vs. 128 channel surface recordings when the standardized weighted swLORETA is used after the use of BEM and SVD to produce a homogeneous lead field similar to that used in MEG (bottom row).

**Figure 4** is from Soler (2010) that compares the localization accuracy between sLORETA and swLORETA and demonstrates not only increased swLORETA

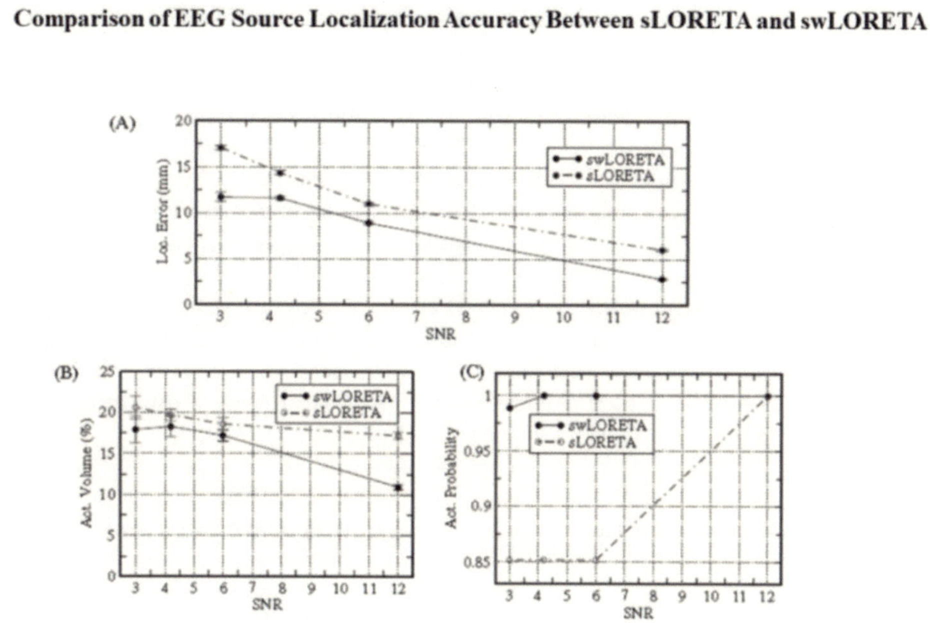


**Figure 4.**
*A comparison of the localization accuracy of sLORETA vs. swLORETA. The X-axis is the signal-to-noise ratio (SNR) and the Y-axis are error measurements. (A) is the localization error in millimeters, (B) is the activation volume as measured by the number of voxels that are 60% or higher than the maximum current location and (C) is the activation probability or how many times out of 300 sources were accurately localized. Reprinted with permission from Soler [26].*

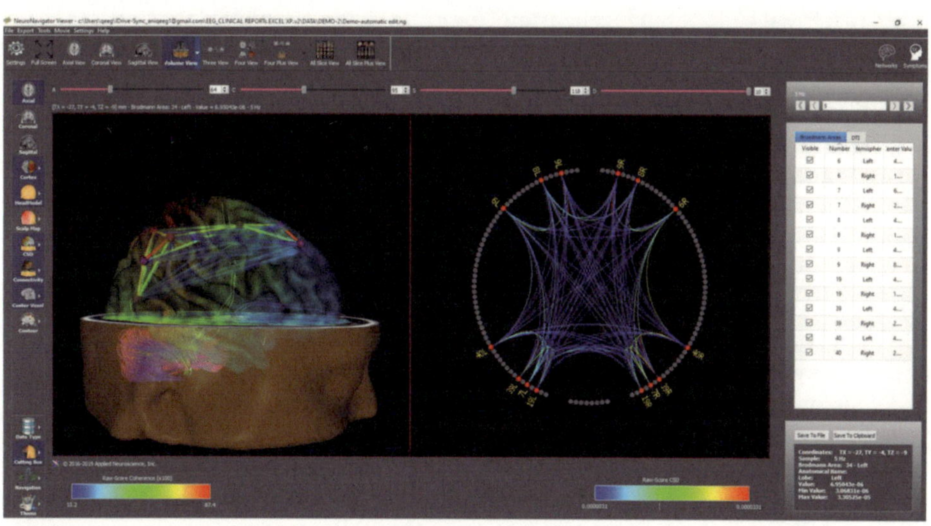

**Figure 5.**
*An example of swLORETA inside of a navigational platform called the NeuroNavigator that allows one to navigate through MRI slices, and the MRI volume to view current sources and functional and effective connectivity. This includes a symptom checklist and brain, networks known to be linked to symptoms based on the human brain mapping program and publications listed in the National Library of medicine (Pubmed). Left is the three-dimensional volume view that includes a semi-transparent cortex, diffusion tensor imaging (DTI) and coherence between the hubs (Brodmann areas) of the dorsal attention network. Right is the two-dimensional "Connectome" of the dorsal attention network selected as one of several possible brain networks as established by human brain mapping fMRI and PET.*

localization accuracy in general but also the ability of swLORETA to image deeper sources than sLORETA. This figure illustrates why swLORETA has the ability to measure deep EEG sources from structures like the cerebellum and red nucleus due to the use of a homogenous lead field, similar to magnetic encephalography (MEG) but with the much more powerful electrical field compared to magnetism.

**Figure 5** is an example of the swLORETA inverse solution inside of a new and powerful viewer called the "NeuroNavigator" that allows one to use a mouse to move through MRI slices in the NIH and Montreal Neurological Institute's template MRI [16, 17]. Talairach coordinates were obtained for every voxel by placing the corresponding Talairach markers onto the anatomical template [18]. The final coordinates of the maxima values (x,y,z, Talairach coordinates) provided for labeling the corresponding brain areas were based on the Talairach atlas. For the definition of cerebellar regions, we used the nomenclature of the MRI Atlas of the Human Cerebellum of Schmahman [19].

## 2. The present: linking symptoms to dysregulated brain hubs and networks

A standard FFT normative database analysis should first be computed in order to identify the electrode locations and EEG features that are most distant from normal and that can be linked to the patient's symptoms and complaints. Linking a subject's symptoms and complaints, e.g., posttraumatic stress disorder, depression, schizophrenia, traumatic brain injury (TBI), etc., to functional localization of networks in the brain is an important objective of those who use a normative database. Similar to a blood bank analysis, the list of deviant or normal measures are given to the clinician as one test among many that are used to help render a diagnosis and to plan treatment. Linking dysregulation of neural activity in localized regions of the brain to known functional localization (for example, left parietal lobe and dyslexia, right frontal and depression, cingulate gyrus and attention deficit, occipital lobes and vision problems) are important to help a trained clinician. Textbooks on functional localization in neurology and psychiatry are available to aid the clinician in learning about the link between a patient's symptoms and different brain regions [27–31]. A link of the anatomical locations and patterns of a patient's deviant z-scores is important in order to derive clinical meaning from the qEEG.

It is the consistency and depth of fMRI, PET, MRI, EEG/MEG studies supported by the human brain mapping project that gave rise to the idea of linking patient symptoms and complaints to brain network hubs and connections in real-time. In 1909 Kobian Brodmann [32] conducted remarkable microscopic studies of human and monkey cadaver brains where he discovered regions of cortical tissue that had a distinct cytoarchitecture of the neurons. Knowing the relationship between structure and function, he concluded that the 44 left and 44 right hemisphere areas or neural clusters must have different functions. Brodmann's work was essentially forgotten until the 1990 human brain mapping program when suddenly PET and fMRI and EEG/MEG confirmed activation of the 88 Brodmann areas by increased blood flow and EEG/MEG source localization related to different functions, e.g., vision and the visual cortex, movement and the motor cortex, etc. [20, 33].

Dynamic hub functional localization in the brain as evidenced by dysregulation of neural populations in Brodmann areas and hemispheres is fundamental to individualized EEG biofeedback. For example, dysregulation is recognized by significantly elevated or reduced power or network measures such as coherence and phase within network hubs and connections of the brain that sub-serve particular functions that can be linked to the patient's symptoms and complaints. The use of z-scores for biofeedback is designed to re-regulate or optimize the homeostasis, neural excitability and network connectivity in particular regions of the brain. Most importantly, the functional localization and linkage to symptoms is based on modern knowledge of brain function as measured by fMRI, PET, penetrating head wounds, strokes and other neurological evidence acquired over the last two centuries [27, 34] also see the Human Brain Mapping database of functional localization at http://hendrix.imm.dtu.dk/services/jerne/brede/index_ext_roots.html). Thousands of published studies in the National Library of Medicine linking symptoms to the brain using fMRI, PET, SPECT, EEG/MEG were made public and available through the internet. In 2009, linking clinical symptoms to dysregulation in brain networks was the backbone of surface and LORETA z-score neurofeedback solely because of the success experienced by patients and advancements in neuroscience.

Once an age-matched qEEG normative database comparison is completed, then one can use a z-score biofeedback program to train patients to move their instantaneous z-scores toward zero or in the direction of the center of the age matched normal population. The absolute value and range of the instantaneous z-scores, while smaller than those obtained using the FFT offline qEEG normative database, are nonetheless valid and capable of being minimized toward zero. An advantage of a z-score biofeedback program is simplification by reducing diverse measures to a single metric, i.e., the metric of a z-score. Thus, as mentioned previously, there is greater standardization and less guesswork about whether to reinforce or suppress coherence or phase differences or power, etc. at a particular location and particular frequency band (see **Figure 1**).

### 2.1 Compensatory vs weak systems

A central concept underlying z-score neurofeedback is distinguishing weak systems from compensatory systems. This distinction was emphasized by Luria [30] and Teuber [35] in their evaluation of patients with penetrating head wounds, strokes and tumors. Modern neuroscience has confirmed the term *neuroplasticity* and neurological compensation in which neural reorganization is measured using EEG, fMRI and PET [36–39]. These studies show that when there is reduced functionality in a given network then reorganization occurs that involves basic neurophysiological mechanisms such as collateral sprouting and compensatory hypertrophy [40]. Specialized networks efficiently process information in coordination with connected modules and hubs in the brain. When there is dysregulation or reduced speed and efficiency of information processing in a subregion or a functional module, then compensatory reorganization often occurs. An example of the role of compensatory reorganization is in an fMRI study of the anxiety network and the role of the frontal lobes in regulation and compensation for dysregulation in subparts of the amygdala [39].

As mentioned previously, the instantaneous z-scores are much smaller than the FFT z-scores in the NeuroGuide software program, which uses the same subjects for the normative database. Smaller z-scores when using the instantaneous z-scores is expected. One should not be surprised by a 50% reduction in JTFA z-scores in comparison to FFT z-scores and this is why it is best to first use 19-channel EEG measures and the highly stable FFT z-scores to link symptoms to functional localization in the brain to the extent possible. Then evaluate the patient's instantaneous z-scores as a therapy or protocol design process before the biofeedback procedure begins. This will allow one to obtain a unique picture of the EEG instantaneous z-scores of each unique patient prior to beginning z-score biofeedback. The clinician must be trained to select which z-scores best match the patient's symptoms and complaints. A general rule for the choice of z-scores to use for biofeedback depends on two factors obtained using a full 19-channel EEG analysis: 1) scalp location(s) linked to the patient's symptoms and complaints, and 2) magnitude of the z-scores. Dysregulation by hyperpolarization produces slowing in the EEG, and dysregulation due to reduced inhibition (hypo-polarization) produces deviations at higher frequencies. The direction of the z-score is much less important than the location(s) of the deviant z-scores and the linkage to the patient's symptoms and complaints.

### 2.2 Z-score neurofeedback publications

In 2006 the first real-time z-score biofeedback method (a DLL or dynamic link library), was developed by Applied Neuroscience, Inc. (ANI) in 2004, and licensed to Brainmaster, Inc. and Thought Technology, LLC. Subsequently, additional EEG biofeedback companies such as Mind Media, Inc., Deymed, Inc. Neurofield, Inc. and EEG Spectrum implemented the ANI real-time z-score DLL. All implementations of live z-score EEG biofeedback share the goal of using standard operant learning methods to modify synapses in brain networks, specifically networks modified by long-term potentiation (LTP) and N-methyl-D-aspartate receptor (NMDA) receptors. Operant conditioning is known to involve changes in the same NMDA receptors that are modified in long term potentiation LTP, and therefore the unifying purpose of z-score biofeedback is to reinforce in the direction of z = 0 of the EEG, which is the statistical center of a group of healthy normal subjects. The normal subjects are a reference just like with blood tests for cholesterol or liver enzymes, etc. that shows deviation from a normative reference database.

As of this date no adverse reactions have been published over the last 13 years nor have adverse reactions been reported by over 3000 clinicians using z-score neurofeedback. This includes six major EEG biofeedback companies, numerous clinicians, Veterans Administration and military medical centers, thousands of patients and over 60 scientific studies. Below is a partial list of scientific studies using z-score EEG biofeedback from 2000 to 2019. Thirty two were published in peer-reviewed journals, 31 were book chapters or International Society for Neurofeedback & Research (ISNR) *NeuroConnections* publications, and four were reviews and or conference presentations. More published research always important and more publications are in progress and will be available in the future. See **Table 1** for a partial list of scientific publications of z-score neurofeedback.

**Table 2** is a summary of the types of patients, clinical disorders and contents of the above z-score neurofeedback publications listed in **Table 1**. Some of the publications included more than one clinical symptom category and some were book chapters with case studies and some were book chapters on z-score methods.

A hypothesized reason that the reinforcement of instantaneous z-scores toward z = 0 is clinically effective is because "chaotic" regimes and extremes of dysregulation are moments of extreme instantaneous z-scores. Reinforcement of "stable" and efficient instances of time results in increased average stability and efficiency in dysregulated nodes and connections in networks linked to symptoms. An analogy is a disruptive child in a school classroom where the teacher gives a reward to the child when the child is quiet and not disruptive. Over time the child will be quiet and more cooperative due to the reinforcement. z-score biofeedback is also consistent with models of homeostatic plasticity in which the learning rule of local inhibitory feedback is increased stability of oscillation around z = 0 [99].

Z-score biofeedback methods are unified by the goal of modifying the brain toward greater homeostasis and inhibiting extreme and unstable states. Z-score biofeedback has its greatest impact on unstable or dysregulated neural systems because unstable systems produce extreme z-scores that are not reinforced and thereby

- Bell et al. [41]
- Collura et al. [42]
- Collura [43–45]
- Collura [46]
- Collura et al. [47, 48]
- Decker et al. [49]
- Duff [50]
- Frey & Koberda [51]
- Foster & Thatcher [52]
- Gluck & Wand [53]
- Groeneveld et al. [54]
- Guan [55]
- Hammer et al. [56]
- Kaur et al. [57]
- Keeser et al. [58]
- Koberda [59]
- Koberda [60–69]
- Koberda & Frey [70, 71]
- Koberda et al. [71–79]
- Krigbaum & Wigton [80]
- Lambos &Williams [81]
- Little et al. [82]
- Lubar [83]
- Pérez-Elvira et al. [84]
- Pérez-Elvira et al. [85]
- Simkin et al. [86]
- Prinsloo et al. [87]
- Smith [88]
- Stark [89]
- Thatcher [90–92]
- Thatcher et al. [93–95]
- Thompson & Thompson [96]
- Wigton [97]
- Wigton & Krigbaum [98]

**Table 1.**
*Partial list of z-score scientific publications.*

| |
|---|
| ADHD = 9 |
| Anxiety = 5 |
| Autism Spectrum Disorder = 2 |
| Dementia = 8 |
| Depression = 3 |
| Epilepsy = 11 |
| Pain = 5 |
| PTSD = 6 |
| Stroke/CVA = 3 |
| TBI =6 |
| Z-score methods = 6 |
| Comparison of the effectiveness of z-score surface/LORETA 19-electrode neurofeedback to standard raw score neurofeedback = 1 |
| Normal subjects in Comparison between fMRI vs. z-Score NFB = 1 |
| 1- to 19-channel surface EEG z-score neurofeedback Publications = 22 |
| LORETA z-score neurofeedback publications = 45 |

**Table 2.**
*Summary of the types of patients, clinical disorders and contents of the z-score neurofeedback publications listed in **Table 1**.*

minimized or extinguished by not being reinforced. The center of the normal population or the ideal instantaneous $z = 0$ is only a momentary ideal state in which homeostatic and balanced systems oscillate around but never achieve perfect $z = 0$ for the entire system. However, on average, unstable neural states that produce large z-score values (e.g., 3 standard deviations or greater) will be minimized and stable neural states that are less than 2 standard deviations will be reinforced. This is the same process at a slower speed that occurs with blood tests. For example, a blood test shows low blood iron compared to the normal population which results in the patient ingesting iron pills, which results in increased blood iron, where z = 0 is the mean of the reference normal population. In the case of z-score biofeedback, the duration and frequency of unstable states or periods of deregulation are reduced as z = 0 is reinforced.

## 2.3 Peak performance

Peak performance has different meanings for different people. A professional golfer who wants to improve his golf game is one thing versus a peak performer traffic controller who wants to do his job better. So being specific about exactly what peak performance is for an individual is critical when dealing with the brain. This is because the brain is the source of all behavior and there are special skills that each person possesses. There is a common misconception that some express by stating: "bringing deviant to normal" is the opposite of what is needed when treating peak performers with z-score EEG biofeedback. This assumption is a bit off because z-score biofeedback is not creating a normal state but rather it is reinforcing stability and efficiency with less network chaos in general. For example, momentary 3 to 6 standard deviations when neurons are not processing information are not reinforced but periods of stability and efficiency less than 2 to 3 standard deviations are reinforced. Operant conditioning reduces the duration and frequency of dysregulation in brain networks and lengthens the average amount of time that groups of neurons are "on-line" and processing information. This represents more neurons and more neural resources available at each instant of time.

No human being is perfect, and a peak performer in golf may not be a peak performer in running or hitting a baseball, etc. What is in common to peak performance are things like efficient memory networks, attention networks, anxiety networks, planning networks, social networks, sensory networks, etc. Therefore, in the hands of a qualified clinician it does no harm to interview a peak performer and ask questions about brain networks like sensation, memory, concentration, attention, anxiety, fear, etc., and then design a z-score protocol to target the brain

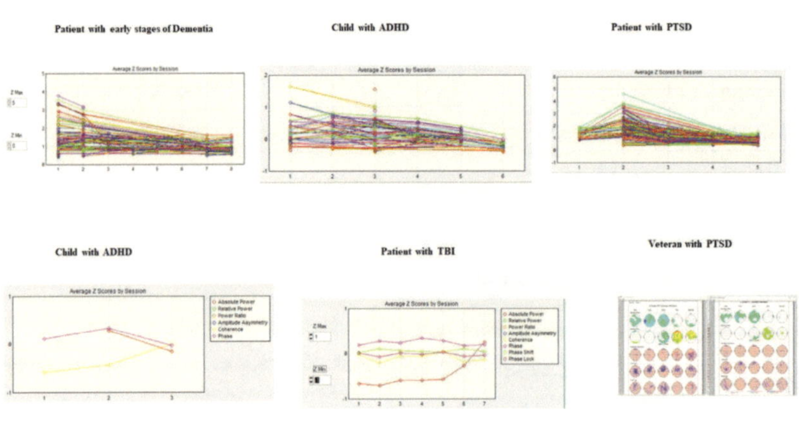


**Figure 6.**
*Examples of changes in z-scores over neurofeedback sessions from different clinicians from their clinical practices from patients with different clinical problems. The Y axis shows z-score values and the X axis shows neurofeedback sessions in six different subjects provided by EEG biofeedback clinicians using surface and/or LORETYA z-score neurofeedback to train patients.*

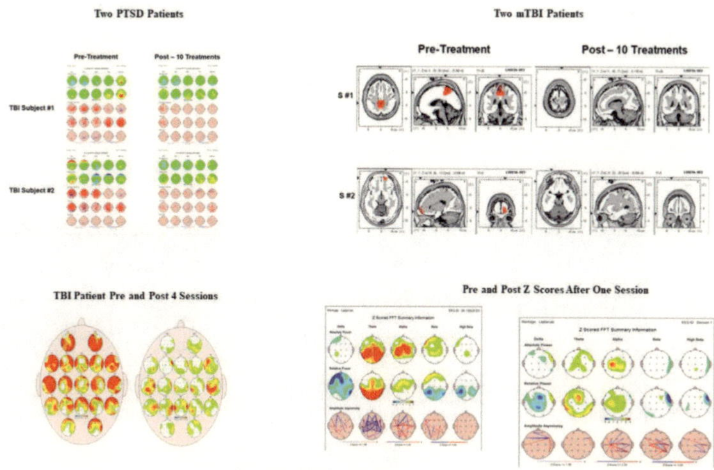


**Figure 7.**
*Examples of reduced z-score values in EEG brain maps in six different subjects in 10 sessions or less from four different clinicians, measured from their clinical practice using EEG z-score neurofeedback.*

regions related to things that the clinician and peak performer believe will help improve their peak performance. It is unlikely that peak performers will be harmed by increased neural stability and increased efficiency in his or her networks. Further, it is important to note that since 2016 numerous EEG biofeedback companies distributed z-score neurofeedback to hundreds of clinicians that have treated thousands of patients and there are no reported examples of a peak performer losing skills or a person with a high IQ becoming less intelligent, etc.

### 2.4 Examples of Z-score change toward Z = 0 over sessions

Reduced z-score values in the direction of z = 0 have been reported in all of the z-score neurofeedback studies published thus far. **Figure 6** are examples of reduced z-scores over sessions shown in a progress chart.

**Figure** 7 are examples of reduced z-scores over sessions shown in scalp surface topographic maps and in LORETA current density maps.

## 3. The future: cerebellar Z-score neurofeedback

Monkey studies of chemically induced Parkinsonism and Cz scalp SMR EEG biofeedback demonstrated reduced Parkinsonism that increased synaptic density and synaptic change in the red nucleus in the SMR group. There were two groups: 1-Dopamine degeneration + SMR and 2- Dopamine degeneration + sham SMR [100, 101].

SMR EEG neurofeedback (12–15 Hz) reduced parkinsonism symptoms were attributed to reinforcing the cerebellum circuits that do not involve dopamine and are a separate and compensatory motor system involved with gait and long movements and legs as one walks. Importantly, the studies of Philippens & Vanwersch [100] and Philippens et al., [101] demonstrated a red nucleus change in synaptic number and organization in the EEG SMR group. The red nucleus is a relay nucleus from the cerebellum the thalamus to motor cortex circuits, with minimal dopamine involved. New advances in EEG Neuroimaging such as swLORETA [12] allow for the evaluation of deep current sources and connectivity from structures such as the cerebellum, red nucleus and the sub-thalamus. This means that in 2019 one can reinforce deep non-dopamine cerebellar and red nucleus circuits that may reduce

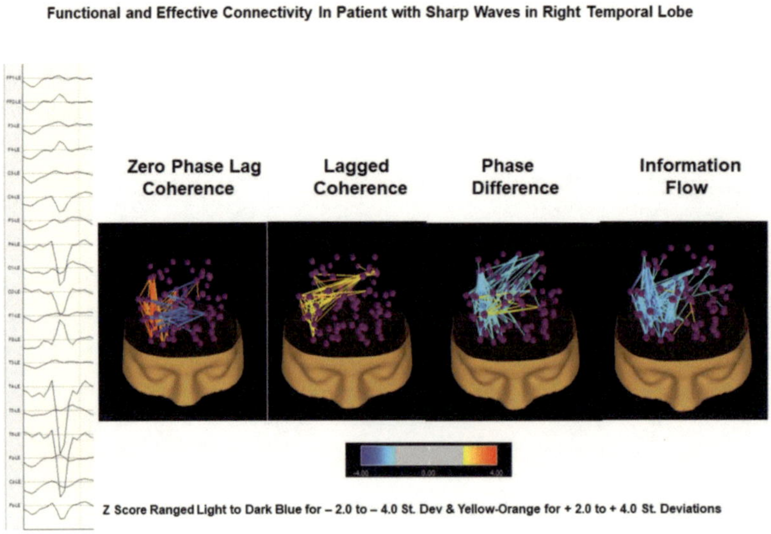


**Figure 8.**
*Example of functional (zero phase lag coherence, lagged coherence and phase difference) and effective connectivity (phase-slope index) between all brain network hubs. This figure illustrates the use of electrical neuroimaging in epilepsy patients where the focal epileptic event is in the right posterior temporal regions. The network analyses allow one to evaluate the local and distant effects on different functional networks and then to evaluate changes over time as a function of treatments.*

Parksonism. As demonstrated by Philippens & Vanwersch [100] and Philippens et al., [101] in monkeys and in studies using the scalp surface EEG SMR which is also directly effects the non-dopamine and non-damaged cerebellar compensatory circuits Thompson & Thompson [102].

Currently we are conducting further verification and validation tests of the cerebellum and red nucleus and subthalamic sources using tDCS and the Rhomberg tests of cerebellar function as well as working with patients with cerebellar infarcts and balance disorders. **Figure 8** is another example of the future application of EEG electrical neuroimaging in the evaluation of epilepsy by measuring both local and long-distance effects of an epileptic focus or sharp waves and the effects of the epileptic event on healthy or non-epileptic networks. A comprehensive evaluation can go beyond localizing the epileptic focus but also understanding the upstream/downstream effects of the focus on distant networks.

The left side of **Figure 9** illustrates some of the anatomical connections of the cerebellum, which is made of three primary lobes: 1) flocculus nodulus (archicerebellum balance and body equilibrium), 2) anterior lobe (paleocerebellar motor execution), and 3) posterior lobe (neocerebeullum – motor plan and coordination). The right side of **Figure 8** illustrates real-time changes in current density produced by clusters of neurons in the various nodes of the cerebellum, which are listed in **Table 3**. EEG Biofeedback starts with real-time auto and cross-spectral measures within and between cerebellar hubs as well as the red nucleus, subthalamus, thalamus and cortex as well as the fully network dynamic as discovered in the 1990s through 2010 by the Human Brain Mapping program, and is continuing today and in 2020.

**Figure 10** shows additional examples of cerebellar EEG sources using swLORETA including real-time functional and effective connectivity and real-time z-score neurofeedback that further confirm the findings of Cebolla et al. [13, 14]. Also, these findings are consistent with the existing scientific literature and long history of the measurement of cerebellar sources from the human scalp EEG (search Pubmed National Library of Medicine database "cerebellar EEG").

*DOI: http://dx.doi.org/10.5772/intechopen.94326*

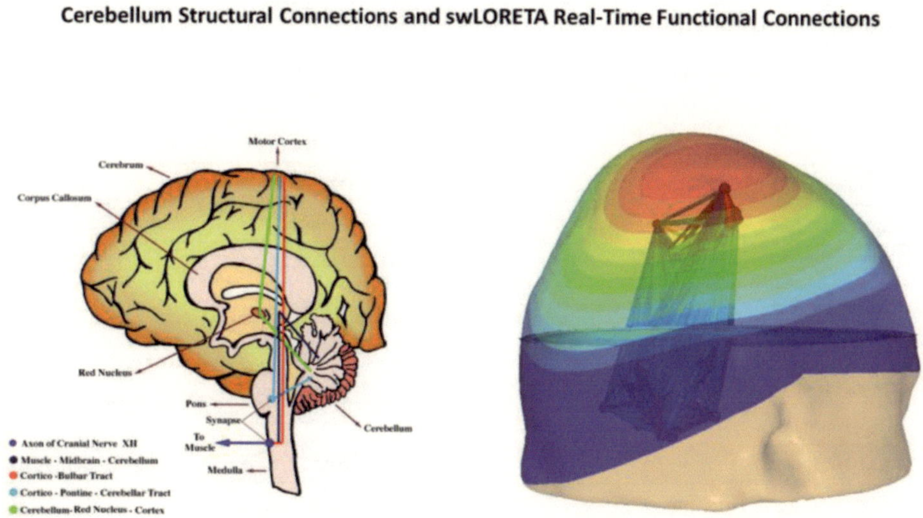


**Figure 9.**
*The image on the left illustrates the anatomical connections of the human cerebellum. On the right is an example of the cerebellum nodes and connections to the sensory-motor cortex using the swLORETA NeuroNavigator (NeuroGuide v. 3.0.7, applied neuroscience, Inc., 2019). Z-scores of the EEG on the scalp surface as well as for functional connectivity between the 13 hubs of the cerebellum, plus the red nucleus, subthalamus and thalamus. See* ***Table 3*** *for a list of the swLORETA neurofeedback protocol options.*

| **Cerebellum EEG biofeedback – menu and protocol selections** | |
|---|---|
| Number | Hemisphere |
| Cerebelum_Crus1 | Left |
| Cerebelum_Crus1 | Right |
| Cerebelum_Crus2 | Left |
| Cerebelum_Crus2 | Right |
| Cerebelum_3 | Left |
| Cerebelum_3 | Right |
| Cerebelum_4_5 | Left |
| Cerebelum_4_5 | Right |
| Cerebelum_6 | Left |
| Cerebelum_6 | Right |
| Cerebelum_7b | Left |
| Cerebelum_7b | Right |
| Cerebelum_8 | Left |
| Cerebelum_8 | Right |
| Cerebelum_9 | Left |
| Cerebelum_9 | Right |
| Cerebelum_10 | Left |
| Cerebelum_10 | Right |
| Vermis_1_2 | Medial |
| Vermis_3 | Medial |
| Vermis_4_5 | Medial |

| **Cerebellum EEG biofeedback – menu and protocol selections** | |
|---|---|
| Vermis_6 | Medial |
| Vermis_7 | Medial |
| Vermis_8 | Medial |
| Vermis_9 | Medial |
| Vermis_10 | Medial |
| Habenula | Left |
| Habenula | Right |
| Sub_Thalamus | Left |
| Sub_Thalamus | Right |
| Red_Nucleus | Left |
| Red_Nucleus | Right |

**Table 3.**
*Shows the wide range of cerebellar sources to select with swLORETA neurofeedback. The cerebellar lobes, vermis, red nucleus, habenula and subthalamus are menu selections for swLORETA neurofeedback based on a patient's symptoms or history such as vertigo, parkinsonism, balance problems.*

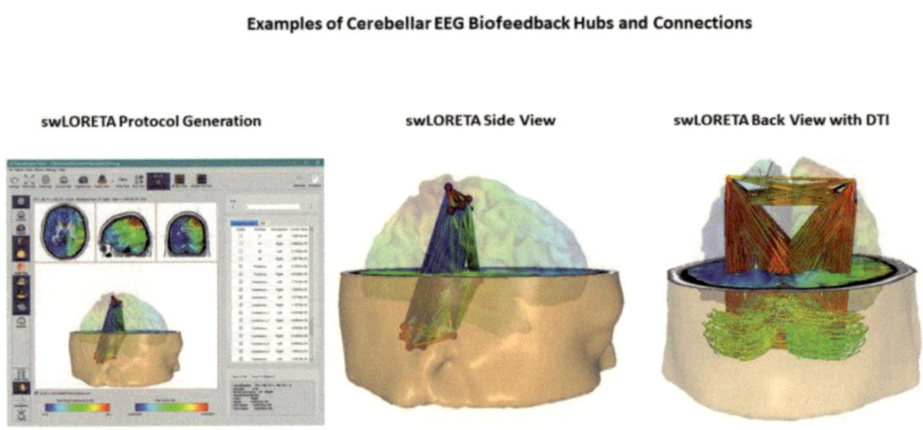


**Figure 10.**
*Examples of swLORETA source localization and functional and effective connectivity between cerebellar sources and the sensory-motor cortex.*

**Table 3** shows some of the cerebellar options to select for cerebellar EEG biofeedback. The cerebellum is made up of three lobes: flocculous nodulous (archicerebellum related to balance and equilibrium), anterior lobe (paleocerebellum related to motor execution), and the phylogenetic more recent posterior lobe (neocerebellum related to motor planning). The vermis is linked primarily to balance and equilibrium, with vermis X as the nodulous part of the flocculous nodulous that receives input from the brainstem vestibular nucleus.

## 4. Conclusions

The universal efficacy of EEG operant conditioning depends on: 1) A time locked external signal to a spontaneously emitted EEG event that predicts a future reward and, 2) Temporal contiguity where there is a limited time window between the emitted EEG event and the feedback signal. A third and important factor is specificity provided by new advances in 3-dimensional electrical neuroimaging of brain networks, i.e., Positive reinforcement of the "weak" node(s) and connections linked to symptoms.

### 4.1 Value of Z score neurofeedback

The use of 19-channel EEG z-score neurofeedback and EEG source localization neurofeedback (LORETA, sLORETA and now swLORETA) can aid in increasing specificity based on the patient's symptoms, informed by the 200 years of neurology as well as the human brain mapping program, beginning in 1990 with the decade of the brain giving rise to three-dimensional fMRI, PET and EEG/MEG assessment of a large number of patients. A unnormalized or raw EEG value fails to provide information about the direction of neurofeedback, i.e., whether to reinforce or to inhibit a given EEG metric. The use of z-score neurofeedback reduces uncertainty and increases simplicity by reducing measures to a single metric of distance from a reference healthy population of age-matched individuals. Reference to a healthy age matched group of individuals helps determine the direction of reinforcement of an EEG event and helps target the weak hubs to reinforce improved regulation and efficiency of brain networks linked to symptoms. The real-time z-score metric identifies outliers or extreme values indicating moments of dysregulation that may be linked to symptoms. The human brain mapping program and the neurological literature, when used with z-scores, aids in identifying dysregulation in the weak hubs and connections of networks linked to symptoms.

This history of z-score neurofeedback, coupled with the science available online, leads toward a modern-day EEG biofeedback protocol that starts with the patient's symptoms followed by an online search of the National Library of Medicine database using the search terms "anxiety brain networks," or "depression brain networks," or "memory brain networks," or "addiction brain networks," etc. depending on the patient's symptoms. This is then followed by the selection Brodmann areas in the hubs and connections of the relevant networks to produce a protocol to reinforce increased stability and efficiency of the networks likely linked to the patient's symptoms.

With the development of improved EEG neuroimaging methods such as weighted swLORETA using over 12,000 MRI voxels and the boundary element method plus the use or a homogeneous lead field improves EEG source localization accuracy closer to that achieved by magnetoencephalography (MEG) at a fraction of the expense. These new developments indicate a bright and promising future for the field of EEG biofeedback by improved source localization accuracy and the ability to link a patient's symptoms to dysregulation in brain networks and connections known to be related to the patient's symptoms. In addition, given these new and inexpensive technologies, the field of EEG biofeedback can expand by helping patients with cerebellar-related problems by enhancing cerebellar compensation in movement disorders like parkinsonism. Parkinsonism strikes approximately 60,000 new patients every year and SMR EEG biofeedback has been shown to reduce the severity of parkinsonism by training the non-dopamine motor system comprising the cerebellum, red nucleus, subthalamus, thalamus and the sensory-motor cortex (SMR = EEG sensory motor rhythms). In the hands of future trained clinicians, physical therapists, chiropractors and ear, nose and throat doctors there will be an increasing use of QEEG to assess and then train toward an improved clinical outcome as demonstrated in human patients [102, 103] as well as in monkeys

[100, 101]. People over age 65 are prone to having balance problems and there are about 40 million Americans older than age 65. Physical therapists measure and use exercises and balance tasks to help patients with balance problems with good success. Nonetheless, it is likely, given the rapid growth of knowledge in neuroscience, that adding a 15- or 20-minute neurofeedback training session that specifically targets the brain's balance system would be effective and harmless.

Education is the key to expanding the applications of EEG biofeedback of all types. Whether z-scores or raw scores, because of the deeper fundamental of self-organization, which is what is accomplished when using EEG biofeedback. Linking symptoms to the patient's brain based on modern science is what drives the future, and because of an absence of serious or debilitating side effects, the FDA has exempted EEG biofeedback companies that use battery powered amplifiers from filing a 510 K form. Caution, however, is always warranted, and education is essential.

## Author details

Robert W. Thatcher[1]*, Carl J. Biver[2], Ernesto Palermero Soler[2], Joel Lubar[2] and J. Lucas Koberda[3]

1 EEG and NeuroImaging Laboratory, Applied Neuroscience Research Institute, St. Petersburg, FL, USA

2 Southeastern Neurofeedback Institute, Pompano Beach, FL, USA

3 Neurology, PL/Brain Enhancement Inc., Tallahassee, FL, USA

*Address all correspondence to: rwthatcher@yahoo.com

IntechOpen

## References

[1] Thatcher R.W., Biver C.J., Soler E.P., Lubar J., Koberda J.L. (2020). New Advances in Electrical Neuroimaging, Brain Networks and Neurofeedback Protocols. J Neurology and Neurobiology, 6(3): dx.doi.org/10.16966/2379-7150.168

[2] Thatcher, R. W. & Lubar, J. F. (2008). History of the scientific standards of QEEG normative databases. In Introduction to QEEG and Neurofeedback: Advanced Theory and Applications, T. Budzinsky, H. Budzinsky, J. Evans & A. Abarbanel (eds)., San Diego, CA: Academic Press.

[3] Thatcher, R. W. (1998). EEG normative databases and EEG biofeedback. Journal of Neurotherapy, 2(4): 8-39.

[4] Thatcher, R. W. (1999) EEG database guided neurotherapy. In: J. R. Evans & A. Abarbanel (Eds.), Introduction to quantitative EEG and neurofeedback, San Diego, CA: Academic Press.

[5] Thatcher, R. W. (2000a). 3-Dimensional EEG biofeedback using LORETA., Society for Neuronal Regulation meeting, Minneapolis, MN.

[6] Thatcher, R. W. (2000b). EEG operant conditioning (biofeedback) and traumatic brain injury, Clinical EEG, 31(1), 38-44.

[7] Balleine BW, Dickinson A (1998) Goal-directed instrumental action: contingency and incentive learning and their cortical substrates. Neuropharmacology 37:407-419. 11.

[8] Schultz W (2006) Behavioral theories and the neurophysiology of reward. Ann Rev Psychol 57: 87-115.

[9] Thatcher RW (2012) Handbook of quantitative electroencephalography and EEG biofeedback. 2nd edition, Anipublishing, St. Petersburg, FL.

[10] Thatcher RW (2016) Handbook of quantitative electroencephalography and EEG biofeedback. 1st edition, Anipublishing, St. Petersburg, FL.

[11] Pascual-Marqui RD, Michel CM, Lehmann D. Low resolution electromagnetic tomography: A new method for localizing electrical activity in the brain. International Journal of Psychophysiology. 1994;18:49-65.

[12] Palmero-Soler, E., Dolan, K., Hadamschek, V. & Tass, P. A. (2007). swLORETA: a novel approach to robust source localization and synchronization tomography. Phys. Med. Biol. **52,** 1783-1800.

[13] Cebolla, A. M., Petieau, M., Dan, B., Balazs, L., McIntyre, J., & Cheron, G. (2016). Cerebellar contribution to visuo-attentional alpha rhythm: Insights from weightlessness. Scientific Reports. 6. 37824. doi:10.1038/srep37824

[14] Cebolla, A.-M., Palmero-Soler, E., Leroy, A. & Cheron, G. (2017) EEG spectral generators involved in motor imagery: A swLORETA Study. Frontiers in Psychology 8. 2133. doi:10.3389/fpsyg.2017.02133

[15] Zanow, F., and Knösche, T. R. (2004). ASA–Advanced Source Analysis of continuous and event-related EEG/MEG signals. Brain Topogr. 16, 287-290. doi: 10.1023/B:BRAT.0000032867.41555.d0

[16] Collins, D. L., Neelin, P., Peters, T. M., and Evans, A. C. (1994). Automatic 3D intersubject registration of MR volumetric data in standardized Talairach space. J. Comput. Assist. Tomogr. 18, 192-205. doi: 10.1097/00004728-199403000-00005

[17] Mazziotta, J. C., Toga, A. W., Evans, A., Fox, P., and Lancaster, J. (1995). A probabilistic atlas of the human brain: theory and rationale for

its development. The International Consortium for Brain Mapping (ICBM). Neuroimage 2, 89-101. doi: 10.1006/nimg.1995.1012

[18] Lancaster JL, Woldorff MG, Parsons LM, et al. Automated Talairach atlas labels for functional brain mapping. Hum Brain Mapp. 2000;10(3):120-131. doi:10.1002/1097-0193(200007)10:3<120::aid-hbm30>3.0.co;2-8

[19] Schmahmann, J. D., Doyon, J., McDonald, D., Holmes, C., Lavoie, K., Hurwitz, A. S., et al. (1999). Three-dimensional MRI atlas of the human cerebellum in proportional stereotaxic space. Neuroimage 10, 233-260. doi: 10.1006/nimg.1999.0459

[20] Thatcher, R.W., Hallet, M., Zeffiro, T., John, E.R. and Huerta, M., Editors. Functional Neuroimaging: Technical Foundations, New York, Academic Press, 1994.

[21] Pascual-Marqui RD. (1999). Review of methods for solving the EEG inverse problem. International Journal of Bioelectromagnetism, 1: 75-86.

[22] Yao, J. and Dewald, J. (2003). Evaluation of Different Cortical Potential Imaging Methods Using Simulated EEG Data. Proceedings of the 25″ Annual International Conference of the IEEE EMBS, Cancun, Mexico - September 17, 2003.

[23] Grech, R., Cassar, T., Muscat, J., Camilleri, K.P., Fabri, S.G., Zervakis, M., Xanthopoulos, P., Sakkalis, V. and Vanrumste, B. (2011). Review on solving the inverse problem in EEG source analysis. Journal of NeuroEngineering and Rehabilitation 2008, 5:25

[24] Songa, J., Daveya, C., Poulsena, C., Luua, P., Turovets, S., Andersona, E., Li, K., & Tucker, D. (2015). EEG source localization: Sensor density and head surface coverage. Journal of Neuroscience Methods. 256: 9-21.

[25] Palmero-Soler, E. and Thatcher, R.W. (2020). Simulation of source localization accuracy using swLORETA and the Neuronavigator (In preparation).

[26] Soler, E. P. (2010). Functional imaging based on swLORETA and phase synchronization. Submitted in partial fulfillment of the requirements for the degree of doctor in science. Available at: https://www.appliedneuroscience.com/PDFs/Ernezto_Soler_2010_Functional_Imaging_based_on_swLORETA.pdf

[27] Brazis, P.W., Masdeu, J.C. & Biller, J. (2007). Localization in Clinical Neurology. Philadelphia, PA: Williams & Wilkins.

[28] Clark, D. L., Boutros, N. N., & Mendez, M.F. (2010). The Brain and Behavior: An Introduction to Behavioral Neuroanatomy, Cambridge, UK:Cambridge University Press.

[29] Tonkonogy, J. N., & Puente, A. E. (2009). Localization of Clinical Syndromes in Neuropsychology and Neuroscience. New York, NY: Springer Publishing.

[30] Luria, A. (1973). The working brain: An introduction to neuropsychology, Baltimore, MD: Penguin Books.

[31] Mesulam, M. (2000). Principles of behavioral and cognitive neurology. Cambridge, MA: Oxford University Press.

[32] Brodmann, V.K., (1909). Localization in the Cerebral Cortex: The Principles of Comparative Localisation in the Cerebral Cortex Based on Cytoarchitectonics, Translated by L. J. Garey, Springer, London, 1994.

[33] Thatcher, R.W., Lyon, G.R., Rumsey, J. and Krasnegor, N. Editors.

Developmental Neuroimaging: Mapping the Development of Brain and Behavior, Academic Press, Florida, 1996.

[34] Heilman, K. M., & Valenstein, E. (1993). Clinical Neuropsychology (3rd ed.)., New York, NY: Oxford University Press.

[35] Teuber, H. L. (1968). Alteration of perception and memory in man. In Weiskrantz, L. (Ed.): Analysis of behavioral change. , 274-328. New York:Harper and Row.

[36] Cabeza, R., Anderson, N. D., Locantore, J. K., McIntosh, A. R. (2007). Aging gracefully: Compensatory brain activity in high-performing older adults. Neuroimage, 17(3).1394-1402.

[37] Chapman, R. M., Porsteinsson, A.P., Gardner, M.N., Mapstone, M., McCrary, J. W., Sandoval, K. and Reilly, L. A. (2013). C145 as a short-latency electrophysiological index of cognitive compensation in Alzheimer's disease. Journal of Alzheimers Disease 33(1). 55-68. doi:10.3233/JAD-2012-120646

[38] Becker, J. T., Mintun, M. A., Aleva, K., Wiseman, M. B., Nichols, T., DeKosky, S. T. (1996). Compensatory reallocation of brain resources supporting verbal episodic memory in Alzheimer's disease. Neurology, 46(3). 692-700.

[39] Etkin, A., Prater, K. E., Schatzberg, A. F., Menon, V., & Greicius, M. D. (2009). Disrupted amygdalar subregion functional connectivity and evidence of a compensatory network in generalized anxiety disorder. Archives of General Psychiatry, 66(12). 1361-1372. doi:10.1001/archgenpsychiatry.2009.104

[40] Geschwind, N., & Galaburda, A.M. (1987). Cerebral lateralization: Biological mechanisms, associations and pathology. Cambridge, MA: MIT Press

[41] Bell, A. N., Moss, D., & Kallmeyer, R. J. (2019). Healing the neurophysiological roots of trauma: A controlled study examining LORETA z-score neurofeedback and HRV biofeedback for chronic PTSD. NeuroRegulation, 6(2), 54-70. doi:10.15540/nr.6.2.54

[42] Collura, T. F. (2008a). Time EEG z-score training: Realities and prospects. In: J. Evans, L. Arbanel, & T. Budsynsky, Quantitative EEG and Neurofeedback, San Diego, CA: Academic Press.

[43] Collura, T. F. (2008b). Whole head normalization using live z-scores for connectivity training. Part 1. NeuroConnections, April, 12-18.

[44] Collura, T. F. (2008c) Whole-head normalization using live z-scores for connectivity training. Part 2. NeuroConnections, July. 9-12.

[45] Collura, T., Guan, J., Tarrent, J., Bailey, J., & Starr, R. (2010). EEG biofeedback case studies using live z-score training and a normative database. Journal of Neurotherapy, 14(1), 22-46.

[46] Collura, T. F. (2009) Practicing with multichannel EEG, DC, and slow cortical potentials, NeuroConnections, January. 35-39.

[47] Collura, T., Thatcher, R., Smith, M. L., Lambos, W., & Stark, C. (2009). EEG biofeedback training using live z-scores and a normative database. Philadelphia, PA: Elsevier.

[48] Collura, T. F. (2009) Practicing with multichannel EEG, DC, and slow cortical potentials, NeuroConnections, January. 35-39.

[49] Decker, S.L. Roberts, A.M. and Green, J.J. (2014). LORETA Neurofeedback in College Students with ADHD. In: RW Thatcher and JF Lubar "Z Score Neurofeedback: Clinical

Applications". Academic Press, San Diego, CA (2014).

[50] Duff, J. (2004) The usefulness of quantitative EEG (QEEG) and neurotherapy in the assessment and treatment of post-concussion syndrome. Clinical EEG and Neuroscience 35(4). 198-209.

[51] Frey, L. C., & Koberda, J. L. (2015) LORETA z-score neurofeedback in patients with medically-refractory epilepsy. Journal of Neurology and Neurobiology. Volume1.1. doi:10.16966/noa.102

[52] Foster, D. S., & Thatcher, R. W. (2014). Surface and LORETA neurofeedback in the treatment of post-traumatic stress disorder and mild traumatic brain injury. In R. W. Thatcher & J. F. Lubar Z Score Neurofeedback: Clinical Applications. San Diego, CA: Academic Press.

[53] Gluck, G., & Wand, P. (2014). LORETA and spec scans: A correlational case series. In R. W. Thatcher & J. F. Lubar Z score neurofeedback: Clinical applications. San Diego, CA: Academic Press.

[54] Groeneveld KM, Mennenga AM, Heidelberg RC, Martin RE, Tittle RK, Meeuwsen KD, Walker LA, White EK. (2019), Z-Score Neurofeedback and Heart Rate Variability Training for Adults and Children with Symptoms of Attention-Deficit/Hyperactivity Disorder: A Retrospective Study. Appl Psychophysiol Biofeedback. 2019 Dec;44(4):291-308. doi: 10.1007/s10484-019-09439-x

[55] Guan, J. (2016). The efficacy of z-score neurofeedback training. In T. F. Collura & J. A. Frederick (Eds.), Handbook of clinical QEEG and neuropathy, 312-325. New York, NY: Routledge.

[56] Hammer, B. U., Colbert, A. P., Brown, K. A., & Ilioi, E. C. (2011). Neurofeedback for insomnia: A pilot study of z-score SMR and individualized protocols. Applied Psychophysiology and Biofeedback, 36(4), 251-264. doi:10.1007 / s10484-011-9165-y

[57] Kaur C, Singh P, Sahni S, Punia C. (2019). Advanced Spatially Specific Neurofeedback for Symptoms of Depression and Its Electroencephalographic Correlates. Altern Ther Health Med., 25(3):54-63.

[58] Keeser, D., Kirsch, V., Rauchmann, B., Stamm, B., Reidler, P., Thatcher, R. W., ... Ertl-Wagner, B. (2014, June). The impact of source-localized EEG phase neurofeedback on brain activity-A double blind placebo controlled study using simultaneously EEG-fMRI. Presentation at Department of Psychiatry, Institute of Clinical Radiology, University of Munich, June 12, Munich, Germany.

[59] Koberda, J. L. (2011). Clinical advantages of quantitative electroencephalogram (QEEG) application in general neurology practice. Neuroscience Letters, 500(Suppl.), e32.

[60] Koberda, J. L. (2012). Autistic Spectrum Disorder as a Potential Target of Z-score LORETA Neurofeedback. The NeuroConnections, 24-25.

[61] Koberda, J. L. (2014a). LORETA z-score neurofeedback in chronic pain and headaches. In: R. W. Thatcher & J. F. Lubar, Z Score Neurofeedback: Clinical Applications. San Diego, CA: Academic Press.

[62] Koberda, J. L. (2014b). Neuromodulation – An emerging therapeutic modality in neurology. Journal of Neurology & Stroke 1(4): 00027. doi:10.15406/jnsk.2014.01.00027

[63] Koberda, J. L. (2014c) QEEG/LORETA electrical imaging in

neuropsychiatry – Diagnosis and treatment implications, In V. Asher-Hansley (Ed.), Advances in Neuroimaging Research, 121-146. Hauppauge, NY: Nova Biomedical Publishing.

[64] Koberda, J. L. (2014d). Therapy of seizures and epilepsy with z-score LORETA neurofeedback. In: R. W. Thatcher & J. F. Lubar (Eds.), Z score neurofeedback: Clinical applications. San Diego, CA: Academic Press.

[65] Koberda, J. L. (2014e) Z-score LORETA neurofeedback as a potential therapy in cognitive dysfunction and dementia. Journal of Psychology & Clinical Psychiatry 1(6): 00037. doi:10.15406/jpcpy.2014.01.00037

[66] Koberda, J. L. (2014f). Z-score LORETA neurofeedback as a potential therapy in depression/anxiety and cognitive dysfunction. In: R. W. Thatcher & J. F. Lubar, Z score neurofeedback: Clinical applications. San Diego, CA: Academic Press.

[67] Koberda, J. L. (2015a) LORETA z-score neurofeedback – Effectiveness in rehabilitation of patients suffering from traumatic brain injury. Journal of Neurology and Neurobiology, 1(4). doi:10.16966/2379-7150.113.

[68] Koberda, JL. (2015b). Application of Z-score LORETA Neurofeedback in therapy of Epilepsy-Editorial-Journal of Neurology and Neurobiology-Vol. 1.1.

[69] Koberda, J. L. (2015c) Traumatic brain injury: Is neurofeedback the best available therapy? Journal of Neurology and Neurobiology 1(3). doi:10.16966/2379- 7150.110

[70] Koberda, J. L., & Frey, L. C. (2015) Z-score LORETA neurofeedback as a potential therapy for patients with seizures and refractory epilepsy – case study. Journal of Neurology and NeurobiologyVolume1.1. doi:10.16966/noa.101

[71] Koberda, JL, & Frey LC. (2015). Application of Z-score LORETA Neurofeedback in Therapy of Epilepsy. Journal of Neurology and Neurobiology-Vol. 1.1.

[72] Koberda, J. L, Hiller, D. S., Jones, B., Moses, A., & Koberda, L. (2012). Application of neurofeedback in general neurology practice. Journal of Neurotherapy, 16(3): 231-234.

[73] Koberda, J. L., Koberda, P., Bienkiewicz, A., Moses, A., Koberda, L. (2013) Pain Management Using 19-Electrode Z-Score LORETA Neurofeedback. Journal of Neurotherapy, 17(3), 179-190.

[74] Koberda, J. L., Koberda, L., Koberda, P., Moses, A., Bienkiewicz, A. (2013) Alzheimer's dementia as a potential target of z-score LORETA 19-electrode neurofeedback. NeuroConnections, Winter, 30-32.

[75] Koberda, J. L., Moses, A., Koberda, L., & Koberda, P. (2012) Cognitive enhancement using 19-electrode z-score neurofeedback. Journal of Neurotherapy 16(3), 224-230 doi:10.1080/10874208.2012.705769

[76] Koberda, J. L., Moses, A., Koberda, P., & Koberda, L. (2012) Comparison of the effectiveness of z-score surface/LORETA 19-electrodes neurofeedback to standard 1-electrode neurofeedback. Proceedings of the 2012 International Society for Neurofeedback and Research (ISNR), Orlando, Florida, Journal of Neurotherapy: Investigations in Neuromodulation, Neurofeedback and Applied Neuroscience, 16:4, 295-315, DOI: 10.1080/10874208.2012.729984

[77] Koberda, J. L., Moses, A., Koberda, P., & Winslow, J. (2014) Cognitive enhancement with LORETA z-score neurofeedback. Association for Applied

Psychophysiology and Biofeedback meeting, Savannah, GA.

[78] Koberda, J. L. & Stodolska-Koberda, U. (2014). Z-score LORETA neurofeedback as a potential rehabilitation modality in patients with CVA. Journal of Neurological Stroke 1(5). 00029. doi:10.15406/jnsk.2014.01.00029

[79] Koberda J.L, Moses A, Koberda P, Koberda L. Comparison of the Effectiveness of Z-Score Surface/LORETA 19-Electrodes Neurofeedback to Standard 1-Electrode Neurofeedback. J. of Neurotherapy, 4, p-302, 2012. rigbaum, G., & Wigton, N. L. (2015). A Methodology of Analysis for Monitoring Treatment Progression with 19-Channel Z-Score Neurofeedback (19ZNF) in a Single-Subject Design. Applied Psychophysiology and Biofeedback. 40(3), 139-149. doi: 10.1007/s10484-015-9274-0

[80] Krigbaum G, Wigton NL. A Methodology of Analysis for Monitoring Treatment Progression with 19-Channel Z-Score Neurofeedback (19ZNF) in a Single-Subject Design. Appl Psychophysiol Biofeedback. 2015 Sep;40(3):139-49. doi: 10.1007/s10484-015-9274-0. PMID: 25777656.

[81] Lambos, W. A., & Williams, R. A. (2014). Treating executive functioning disorders using LORETA z-scored EEG biofeedback. In R. W. Thatcher & J. F. Lubar, Z score neurofeedback: Clinical applications. San Diego, CA: Academic Press.

[82] Little, R. M., Bendixsen, B. H., & Abbey, R. D. (2014). 19 channel z-score training for learning disorders and executive functioning. In R. W. Thatcher & J. F. Lubar, Z score neurofeedback: Clinical applications. San Diego, CA: Academic Press.

[83] Lubar, J. L. (2014). Optimal procedures in z score neurofeedback: Strategies for maximizing learning for surface and LORETA neurofeedback. In: R. W. Thatcher & J. F. Lubar, Z score neurofeedback: Clinical applications. San Diego, CA: Academic Press.

[84] Pérez-Elvira, R., Carrobles, J. A., López Bote, D. J., & Oltra-Cucarella, J. (2019). Efficacy of live z-score neurofeedback training for chronic insomnia: A single-case study. NeuroRegulation, 6(2), 93-101. https://doi.org/10.15540/nr.6.2.93

[85] Pérez-Elvira, R., López Bote, D. J., Guarino, S., Agudo Juan, M., De León, R. J., Feiner, T., & Perez, B. (2018). Neurometric results of a case series using live z-scores neurofeedback. International Journal of Psychophysiology, 131, S139–S140. https://doi.org/10.1016/j.ijpsycho.2018.07.375

[86] Simkin D. R., Thatcher R. W., & Lubar J., (2014). Quantitative EEG and neurofeedback in children and adolescents: Anxiety disorders, depressive disorders, comorbid addiction and attention-deficit/hyperactivity disorder, and brain injury, Child and Adolescent Psychiatriatric Clinics of North America 23(3), 427-464, doi:10.1016/j.chc.2014.03.001

[87] Prinsloo, S., Rosenthal, D. I., Lyle, R., Garcia, S. M., Gabel-Zepeda, S., Cannon, R., Cohen, L. (2019). Exploratory study of low resolution electromagnetic tomography (LORETA) real-time z-score feedback in the treatment of pain in patients with head and neck cancer. Brain Topography. 32(2), 283-285. doi:10.1007/s10548-018-0686-z.

[88] Smith, M. L. (2008). Case study: Jack. NeurosConnections, April.

[89] Stark, C. R. (2008). Consistent dynamic Z-score patterns observed during Z-score training sessions – Robust among several clients

and through time for each client. NeuroConnections, April.

[90] Thatcher, R. W. (2000a). 3-Dimensional EEG biofeedback using LORETA., Society for Neuronal Regulation meeting, Minneapolis, MN.

[91] Thatcher, R. W. (2010). LORETA z-score biofeedback. NeuroConnections, December, 14-17.

[92] Thatcher, R. W. (2012): Latest developments in live z-score training: Symptom check list, phase reset, and LORETA z-score biofeedback, Journal of Neurotherapy: Investigations in Neuromodulation, Neurofeedback and Applied Neuroscience, 17(1), 69-87.

[93] Thatcher, R. W., North, D. M., & Biver, C. J. (2014a). Technical foundations of z score neurofeedback. In R. W. Thatcher & J. F. Lubar Z Score Neurofeedback: Clinical Applications. San Diego, CA: Academic Press.

[94] Thatcher, R. W., North, D. M., & Biver, C. J. (2014b). Network connectivity and LORETA z score NFB. In R. W. Thatcher & J. F. Lubar (eds.), Z Score Neurofeedback: Clinical Applications. San Diego, CA: Academic Press.

[95] Thatcher, R. W., North, D. M., & Biver, C. J. (2014c). BrainSurfer 3-dimensional z score brain-computer-interface. In: R. W. Thatcher & J. F. Lubar Z Score Neurofeedback: Clinical Applications. San Diego, CA: Academic Press.

[96] Thompson, M., Thompson, L., & Reid-Chung, A. (2014). Combining LORETA z-score neurofeedback with heart rate variability training. In R. W. Thatcher & J. F. Lubar (eds.) Z Score Neurofeedback: Clinical Applications, San Diego, CA: Academic Press.

[97] Wigton, N. L. (2013) Clinical perspectives of 19-channel z-score neurofeedback: Benefits and limitations, Journal of Neurotherapy: Investigations in Neuromodulation, Neurofeedback and Applied Neuroscience, 17(4), 259-264.

[98] Wigton, N. L., & Krigbaum, G. (2015). Attention, executive function, behavior, and electrocortical function, significantly improved with 19-channel z-score neurofeedback in a clinical setting: A pilot study. Journal of Attention Disorders . pii: 1087054715577135. [Epub ahead of print]. doi:10.1177/1087054715577135

[99] Hellyer, P.J., Jachs, B., Clopath, C. and Leech, R. (2016). Local inhibitory plasticity tunes macroscopic brain dynamics and allows the emergence of functional brain networks. Neuroimage, 124:85-95

[100] Philippens, I. H. & Vanwersch, R. A. (2010). Neurofeedback training on sensorimotor rhythm in marmoset monkeys. NeuroReport 21, 328-332, https://doi.org/10.1097/WNR.0b013e3283360ba8

[101] Phillippens, I. H., Wubben, A., Frank, S. K., Hofman, S., & Langermans, J. A. (2019). Involvement of the red nucleus in the compensation of parkinsonism may explain why primates can develop stable Parkinson's disease. Scientific Reports, 9, 880 doi:10.1038/s41598-018-37381-1

[102] Thompson, M., & Thompson, L. (2002). Biofeedback for movement disorders (dystonia with Parkinson's disease): Theory and preliminary results. Journal of Neurotherapy 6(4), 51-70. doi:10.1300/J184v06n04_06

[103] Philippens, I. H. C., Wubben, J. A., Vanwersch, R. A. P., Estevao, D. L. & Tass, P. A. (2017). Sensorimotor rhythm neurofeedback as adjunct therapy for Parkinson's disease. Ann Clin Transl Neurol 4, 585-590, https://doi.org/10.1002/acn3.434

Section 3

# Neuromeditation

Chapter 3

# Neuromeditation: The Science and Practice of Combining Neurofeedback and Meditation for Improved Mental Health

*Jeff Tarrant*


## Abstract

Beginning meditators often complain that they do not know if they are "doing it right" or give up before realizing significant benefits. Advanced meditators often reach a plateau and struggle to reach "the next level" of their practice. Modern researchers and practitioners are finding a possible new solution to these challenges by using EEG biofeedback to increase awareness of subtle states of consciousness and speed the learning process. By tracking brainwave activity in specific regions of the brain, we can tell if someone is focused or relaxed. We can tell if the mind is wandering, if they are engaged in body-based emotions, or if they have entered a space of internal quiet. By monitoring this activity and connecting it directly to the intent of the meditation, it is possible to help meditators learn to quickly enter a desired state of consciousness and maintain this state for increasing periods of time. This chapter will describe the early research conducted in this area along with an original case study conducted by the author. In addition, the author will describe the way this technology is being used as a treatment intervention for ADHD, anxiety, depression, and PTSD.


**Keywords:** neuromeditation, neurofeedback, meditation, mindfulness, EEG biofeedback, neurotherapy

## 1. Introduction

There is abundant evidence that the practice of meditation can lead to improvements in an array of physical and mental health concerns [1]. Not surprisingly, this has led to increasing acceptance of these practices in Western societies. In fact, a recent survey found that three of the top four reasons for starting a meditation practice related to improved mental health or affect management [2]. Despite the increased interest in secular-based meditation programs designed to reduce stress or improve mental well-being (e.g., Mindfulness Based Stress Reduction, Mindfulness Based Cognitive Therapy), many people continue to find it difficult to begin or maintain a consistent practice, giving up before they realize any significant benefit.

Researchers, therapists, and meditation coaches are finding a possible new solution to these challenges by using EEG biofeedback to increase awareness of subtle states of consciousness and speed the meditation learning process [3–5].

## 2. Neurofeedback and meditation

### 2.1 Neurofeedback explained

Neurofeedback, sometimes referred to as EEG biofeedback, involves measuring brain wave activity through an electroencephalogram (EEG) and using that information to help the brain understand and modify its processes [6]. Because the raw EEG is a complex signal containing a wide range of frequencies, such data are typically filtered and organized into clusters, called bins. For example, alpha brainwaves are typically identified as the activity occurring between 8 and 12 Hz, while beta brainwaves can be identified as the activity between 15 and 25 Hz. The amount of activity recorded in each of these EEG clusters is measured in microvolts (mv). So, for each electrode used in a recording, it is possible to identify an average amount of power (mv) for each of the specified EEG bands (e.g., delta, theta, alpha, etc., see **Figure 1**).

Once the EEG signal has been quantified, it is connected to computer-based audio and visual signals (feedback) that change in response to the EEG patterns. In this way it is possible to create a pleasant signal that occurs when the brain moves in the desired direction and remove the signal when the brain moves in an undesired direction. With repeated exposure to this process the brain can learn to become more flexible and adaptive, shifting out of rigid states that may be connected to particular concerns, such as ADHD or epilepsy [7]. The neurofeedback provider is trained to understand EEG patterns in relation to specific concerns and is able to create individualized programs for each client based on their goals and needs.

### 2.2 EEG patterns and states of consciousness

EEG bands are clusters of frequencies, organized into groups based loosely on their shape and function. These bands vary in their definition depending on the specific researcher or clinician. Typical EEG band ranges and descriptions follow.

Delta waves (0–4 Hz) are the slowest brainwaves. When they are dominant, the person is most likely asleep. While we always produce delta activity, if it increases significantly in relation to the other EEG bands, it will be very difficult to maintain any sort of alert consciousness. Its function appears to be mostly related to rest and regeneration.

Theta waves (4–8 Hz) are also considered slow brainwaves, although a bit faster than Delta. This band tends to increase during memory retrieval, creative thinking,

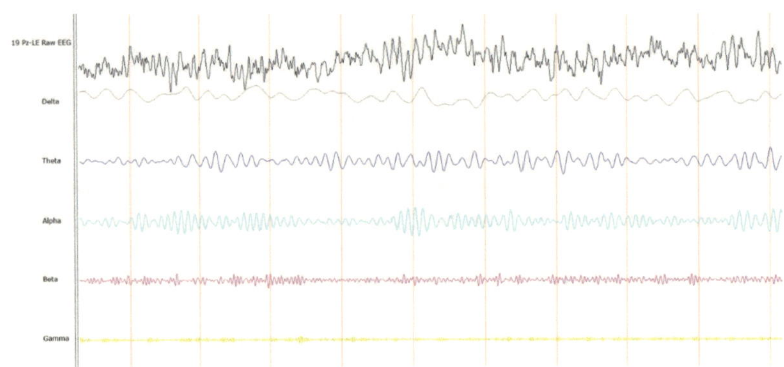


**Figure 1.**
*Raw, delta, theta, alpha, beta, and gamma EEG bands.*

and the "twilight state" just before falling asleep. Theta waves are often associated with the mind being in a more receptive state, such as might occur during hypnosis.

Alpha waves (8–12 Hz) represent our "idle" speed, between the slow and fast waves. Alpha is generally associated with being relaxed and internally focused, hence its historical connection to meditation. Alpha activity tends to increase in the absence of stimulation and is frequently viewed as an inverse indication of activation.

Beta waves (12–30 Hz) are fast waves and associated with activation and arousal. When beta increases, it is likely that the person is engaged in thinking, planning, worrying, or some other active state.

Gamma waves (35–50 Hz) are associated with a very sharp focus and feelings of creativity and insight. Increases in this activity are often observed during high-level information processing or a more effortless, but complex form of understanding, such as occur in a flow state.

### 2.3 The meditating brain

Contrary to popular belief, there is no single EEG pattern associated with meditation. This is largely due to the fact that there are many different approaches to meditation with distinctly different ways of directing attention. In addition, each style of meditation impacts specific regions of the brain. For example many forms of a Focus or Concentration meditation practice result in activation of the frontal lobes while simultaneously showing a de-activation of regions in the back of the brain.

Based on reviews of the EEG meditation literature, most researchers agree that there are four basic styles of meditation defined by how attention is directed, the intention of the meditator, which brainwaves are involved, and in which brain regions [5, 8, 9]. These four styles can be described as follows:

**Focus:** Meditative practices with this emphasis involve sustaining attention on a single object, such as the breath or a mantra. When the mind wanders, the task is to recognize this and return to the original point of focus. Regardless of the specific target, practices in this category require sustaining attention and minimizing mind wandering. Consequently, neuromeditation approaches with this style in mind must monitor specific regions of the frontal lobe (sustaining attention) and the Default Mode Network (DMN; mind wandering). The goal is to keep the attention circuits activated without becoming caught in self-referencing narratives.

**Mindfulness:** While the term mindfulness has been popularized to refer to a range of practices, it is being used here to describe meditation styles that require the meditator to shift into an observer state of awareness, gently watching thoughts, feelings, and bodily sensations without attachment. It is a present moment awareness without attempts to control, analyze, or judge the experience. These practices also quiet down the Default Mode Network while simultaneously activating the Salience Network, which directs attention toward what is important in the moment.

**Open heart:** These practices involve activating a positive feeling state and directing those feelings toward self or others. Practices such as lovingkindness, compassion, gratitude, and forgiveness-based meditations fit in this category. These practices activate attention networks and brain regions associated with empathy and emotional processing.

**Quiet mind:** Practices in this category represent the stereotype of meditation. This is a state in which internal chatter has been reduced to a minimum. Sometimes it is described as a feeling of spaciousness or emptiness. This state is common in traditions like Zen or Transcendental Meditation (TM). Not surprisingly, the brain patterns connected to these practices show a significant quieting of many regions of the brain, including the Default Mode Network and language centers.

## 3. Understanding Neuromeditation

Simply put, neuromeditation is the combination of meditation with neurofeedback. By monitoring brainwave activity in specific regions of the brain, it is possible to determine if someone is focused or relaxed, if the mind is wandering, if they are engaged in body-based emotions, or if they have entered a space of internal quietude. By tracking this activity in real-time and connecting it directly to the intent of the meditation, it is possible to help meditators learn to quickly enter a desired state of consciousness and maintain this state for increasing periods of time, increasing the impact and effectiveness.

### 3.1 History of Neuromeditation

The practice of combining neurofeedback with meditation is not new. In fact, many of the pioneers in the field of neurofeedback were motivated by the desire to enhance their meditation practice or explore states of consciousness. Based on the research of the time, meditation was primarily associated with increased Alpha brainwave activity, particularly in the Occipital and Parietal regions of the brain [10, 11]. As a result, rewarding increases in Alpha amplitude in these regions became the "go-to" approach to neuromeditation for many years [11]. Because increases in the Alpha band are generally related to an inhibition of mental activity [12], this approach was useful in achieving Quiet Mind meditative states, consistent with certain TM or Zen practices [4, 8]. It was also found that protocols designed to reward increases in Alpha amplitude frequently resulted in decreased anxiety, feelings of relaxation, and positive emotions, providing mental health benefits to those suffering from chronic stress or anxiety [13, 14].

### 3.2 Neuromeditation example

By placing an EEG electrode near the back of the head, it is possible to gain information about the state of the brain's Default Mode Network (DMN). The DMN is a vast network with its primary hub located in the Posterior Cingulate Cortex (PCC) of the Parietal lobes (see **Figure 2**), just beneath electrode site PZ (see **Figure 3**). This area of the brain becomes active when a person is involved in self-referential thought [16]. Basically, any thought you have that relates back to your view of self or your connection to the world will involve the DMN. Not surprisingly, activity in the DMN (and PCC) is connected to mind wandering during a meditative practice [17]. Essentially, if you are not fully engaged in the intention of the meditation, you are likely thinking about yourself or something related to yourself. By quieting the DMN, represented by increases in Alpha amplitude or decreases in Beta or Gamma, it is possible to move beyond the typical "story-telling" tendencies of the mind and tap into an internalized, peaceful state of consciousness.

When the Alpha brainwave patterns increase in this region, it is likely that the internal state is more relaxed and the mental activity is inhibited. When Alpha is lower, there is likely to be more analyzing, judging, comparing, remembering, or planning. By establishing a threshold marker in the neurofeedback software, it is possible to identify when the Alpha is "high" or "low."

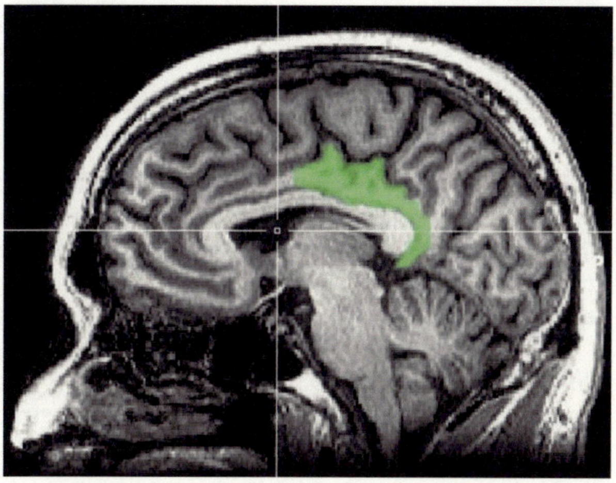

**Figure 2.**
*Sagittal MRI slice with highlighting indicating location of the posterior cingulate cortex [15].*

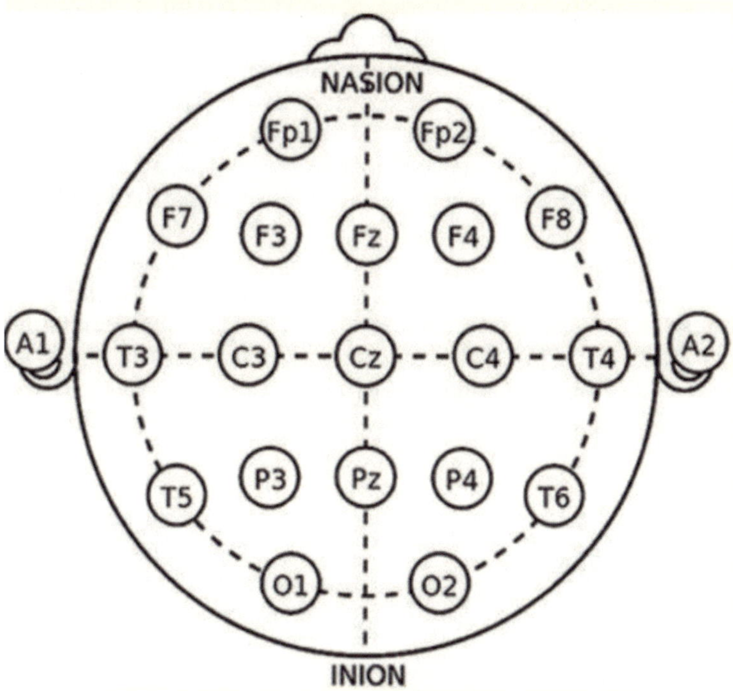


**Figure 3.**
*International 10-20 EEG electrode map.*

When the Alpha activity increases and moves above the threshold marker, the meditator receives some form of pre-determined feedback, letting them know they are on the right track. The feedback used for meditation is typically some form of audio signal or change in music volume designed to provide information without disrupting the meditative state. When the mind wanders, the Alpha drops, signaling a change in the audio signal (e.g., decrease in volume). This provides direct and nearly immediate feedback to the meditator, allowing them to refine their internal awareness.

It should be noted that the example provided above is an over-simplification of the process, but offered for the purposes of illustration.

### 3.3 The research evidence for Neuromeditation

While the real-world applications of neuromeditation have been explored in the neurofeedback community for many years, there have been only a few studies demonstrating the power of this approach in the lab. The first study to examine the feasibility of neurofeedback for meditation used real-time fMRI data to examine the subjective experience of meditators when the Posterior Cingulate Cortex (PCC) was active vs. quiet. Rather than measuring EEG activity, this study examined blood flow, which is an indication of activation [18].

The meditators in this study reported experiences of being "distracted," "interpreting," "controlling," and "efforting," when the PCC was active. In contrast, they reported experiences of "concentration," "undistracted awareness," "effortless doing," and "observing sensory experience" when the PCC was deactivated. A follow-up study provided feedback to meditators about the activity level of the (PCC) during a Focus style of meditation [19]. Both meditators and non-meditators reported a significant relationship between activation of the PCC and mind wandering as well as deactivation of the PCC and focused attention.

The researchers also found that experienced meditators (but not novice meditators) were able to intentionally decrease activation of the PCC through the use of the feedback [18]. In another study using EEG neurofeedback, van Lutterveld, et al., found that both novice and experienced meditators were able to control the experience of effortless awareness in connection with a feedback signal indicating decreased PCC activation [20].

More recently, eight sessions of neurofeedback-enhanced meditation were compared to a control group that received sham neurofeedback [21]. Rather than focusing on the PCC, these researchers rewarded increases of frontal midline theta brainwaves (FM Theta). FM Theta are slow oscillations, between 4 and 7 Hz that are generated in the Anterior Cingulate Cortex (ACC; see **Figure 4**). Theta originating in the ACC often increases in power during a variety of cognitive processes

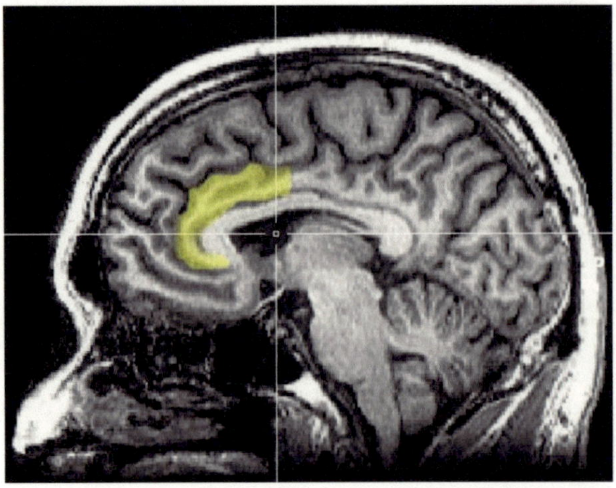

**Figure 4.**
*Sagittal MRI slice with highlighting indicating location of the anterior cingulate cortex [15].*

that require attention, focus, or emotional processing [22–25]. Several studies have found a correlation between increased FM Theta and focused attention meditation practices [26, 27]. Results showed that the experimental group was not only able to significantly increase FM Theta, but also improved performance on a working memory task [21].

Video 1 (https://youtu.be/OxMdYj2Jq4Y; [28]) provides a demonstration and explanation of a neuromeditation protocol that examines increases of FM Theta while simultaneously monitoring (de)activation of the DMN.

## 4. Neuromeditation for mental health

While most work in the field of neuromeditation has been devoted to enhancing or advancing the development of specific meditative states, clinicians and researchers are also beginning to explore this strategy as an intervention to improve mental health and cognitive functioning. Because both meditation and neurofeedback have independently been found to be effective in the treatment of a variety of mental health concerns, it is logical to combine them to target specific outcomes [3, 5]. For example, in the study cited above subjects receiving eight sessions of focus neuromeditation significantly improved their performance on a working memory task, while the control group did not [21]. These outcomes make sense given that regions of the brain involved in working memory are exercised during Focus meditation practices. With this logic, it is possible to identify which styles of meditation might be best suited for particular outcomes.

Focus practices with an emphasis on sustaining attention on a single object, activate the frontal lobes, making it an ideal practice for improving functions related to attention, memory, or other executive functions [8, 29, 30]. Consequently, this might be the most beneficial practice for someone with ADHD, cognitive decline, or traumatic brain injury.

Mindfulness, which involves a much more relaxed, observing form of attention may be best suited for managing stress and anxiety [31]. A key component of mindfulness practices involves non-attachment and learning to let go [32], key elements involved in managing stress and anxiety. Not surprisingly, these practices have been shown to reduce activation of the Amygdala, a key brain region involved in the fight or flight response [33].

Open Heart practices, such as lovingkindness-compassion, and gratitude engage positive feeling states, increasing empathy, perspective-taking, and the experience of joy and appreciation [8, 30]. These practices can be helpful for those dealing with resentment, unresolved grief, anger management, or depression.

Practices in the Quiet Mind category result in a reduction of self-talk, leading to the experience of spaciousness or emptiness [8, 34]. Because these practices essentially involve interrupting the "normal" process of "selfing," they can be helpful for concerns connected to a distorted or inaccurate perception of self, which includes most mental health concerns.

While the four styles can certainly serve as a guide for matching a person to the ideal meditation practice, there are often levels of nuance that require assessment and direction from a trained mental health professional. This is particularly true for clients engaging in neuromeditation with unresolved trauma. In addition, we have found that EEG guided meditation is most effective when it is individualized and includes meditation coaching. The case study below will demonstrate this approach.

### 4.1 Case study

#### *4.1.1 Background*

B.A. is a 39-year-old, Caucasian woman with a mental health history of anxiety, eating disordered behaviors, and post-traumatic stress disorder (PTSD). The PTSD relates to a car accident and childhood sexual trauma.*.

B.A. began working with yogic practices in her 20s. She noted that she would frequently experience strong emotions such as grief and anger while holding certain poses which led to some resistance to these practices. She began practicing Transcendental Meditation during her late 20s, but never felt confident in this practice. She described mostly engaging in brief practices and struggles with judging herself.

She identified three concerns she hoped to address through her neuromeditation practice; these included: a tendency to be hyper-critical of self and others, feeling overwhelmed and sensitive to sound, and a desire to feel more grounded-to slow down. These were all rated as moderate concerns. In elaboration of item 3, B.A. noted that she is "very much in her head" and often feels disconnected from her body.

Known barriers to expanding her current meditation practice include time, internal resistance, the critical mind, and a tendency to "leave her body" when she begins to relax.

*Identifying information related to this client has been altered to protect their identity. In addition, the client has given permission for their case to be shared in this format.

#### *4.1.2 Assessment results*

Results of the neuromeditation Styles Inventory [35] indicated that her concerns most closely matched the Quiet Mind style of meditation. Elevated scores on the New Mind Cognitive Emotional Checklist (CEC) Symptom Checklist [36] indicated concerns with memory, sensitivity to light and sound, feeling "spacey" or "out of my body," and thinking obsessively. In addition to the above, a Quantitative EEG assessment provided a comparison of her baseline EEG activity to a clinical database. Using a Laplacian reference for the eyes closed data set and analyzing it through qEEG Pro [37], the most striking feature was increased absolute power across all EEG bands in similar brain regions (see **Figure 5**).

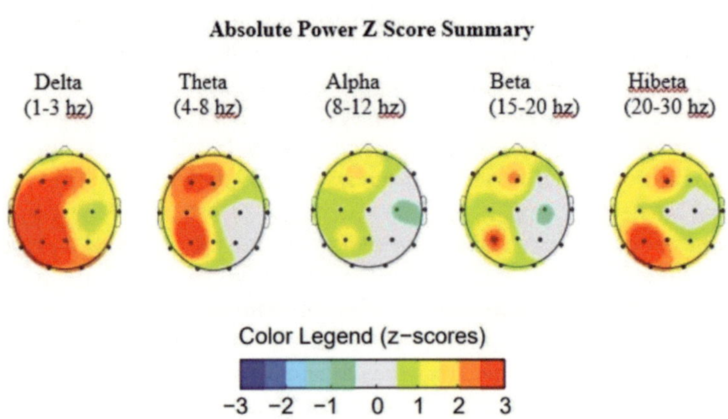


**Figure 5.**
*Absolute power across EEG bands compared to qEEG Pro clinical database.*

The EEG analysis indicates that Delta (1–3 Hz) and Theta (4–8) activity were elevated in left frontal and parietal regions. These same regions also demonstrated elevated Beta (15–20 Hz) and Hibeta (20–30 Hz) which appeared more localized at FZ and P3. Alpha activity was largely within normal limits.

The combination of excessive slow and fast activity in similar regions with average alpha activity suggests that this pattern may be related to the PTSD concerns noted in the interview. Specifically, the increased slow activity may be connected to the tendency to dissociate. This pattern could also be related to some of the memory, attention, and impulsivity concerns noted in the CEC. The elevated fast activity may be related to tendencies toward anxiety as well as sensory sensitivity.

#### *4.1.3 Individualized approach*

Based on the information gathered, a Mindfulness meditation protocol was identified as the best match for her concerns and background. Specifically, this protocol would reward increased activation of the right Insula and deactivation of the PCC. Activity in the Right Insula is a common finding in Mindfulness practices and relates to interoception, emotional self-awareness [38], and metacognitive awareness [39]. The right Insula was highlighted as it tends to be more connected to a felt sense of the body and may be helpful in feeling more grounded (one of B.A.'s goals). The reduction of activity in the PCC will require a limit on cognitive processing such as analysis, comparison, or creating a narrative about the experience. This aspect of the protocol addresses concerns around "thinking obsessively."

#### *4.1.4 Session results*

##### *4.1.4.1 Session 1*

B.A. was initially instructed to simply observe whatever she notices in her body as sensation without any interpretation or internal dialog about these observations. Near the 11-min mark of the meditation, B.A. came out of her meditation and commented that she was having trouble feeling into her body and tends to "disappear," feeling nothing. The therapist provided grounding skills training including rubbing her fingers together, or tapping her fingers to her thumb to create a tactile sensation. B.A. attempted this for approximately 6 min and stopped the session again. She noted that she was struggling with this practice. After additional discussion, B.A. agreed to try focusing on the heart by imagining breathing into and out of the heart, attending to any sensations in that area. After the session, she noted that this seemed to work better for her and was assigned as homework. **Figure 6** below shows an analysis of the session in the neuromeditation Report Writer. Each of the time segments described above were identified in the EEG record for comparison. The scores indicate the percent of time that she was able to keep the identified EEG activity in the desired direction. It is clear from examining the right Insula, PCC, and the combined success, that B.A. was much more skilled at reducing activity in the PCC than in increasing activity in the right Insula. This is consistent with her report that she "felt nothing" and tended to "disappear." This was also consistent with her history of practicing TM meditation, which falls into the Quiet Mind category.

##### *4.1.4.2 Session 2*

The session began with a discussion of her home practice during the past week (without neurofeedback assistance). We explored her tendency to try too hard, become impatient with herself and judge "success." B.A. was encouraged to relax

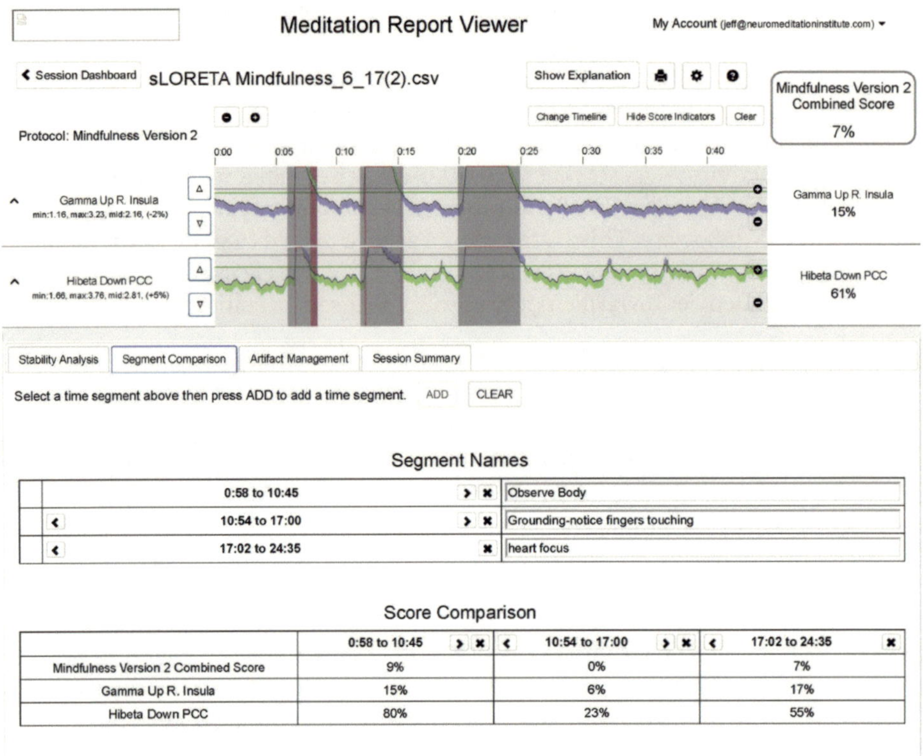


**Figure 6.**
*Analysis of session 1 in the NeuroMeditation report writer. Scores reflect the percent of time that EEG criterion was met for specified meditation segments.*

her goals and expectations for the practice, allowing time to find the meditative space. B.A. was able to connect to sensations in her throat during the session. The time periods of the session where this occurred were easily identified with increased gamma activity in the *R. insula*. While this approach appeared successful, B.A. reported feeling somewhat "panicky" near the end of the session. B.A. indicated that focusing on the throat was causing some trauma related feelings and memories connected to a history of sexual abuse. We discussed her reaction briefly, inviting her to change the focus of her meditation, engage with eyes open, use a variety of grounding tools, or titrating the experience to maintain a feeling of safety.

#### *4.1.4.3 Session 3*

B.A. noted that the tendency to dissociate is so strong that it requires a lot of energy and effort to stay present. To encourage present focused awareness, B.A. began coaching herself internally, reminding herself that she is safe, noting her process and experience. While this strategy helped B.A. to stay in her body without dissociating, the internal narrative caused the activity in the PCC to increase. Consequently, the EEG analysis for this session showed increased success with the *R. insula*, but decreased success with the PCC.

#### *4.1.4.4 Session 4*

B.A. noted more spontaneous experiences of mindfulness outside of session and fewer dissociative moments. During the session we altered the instructions and

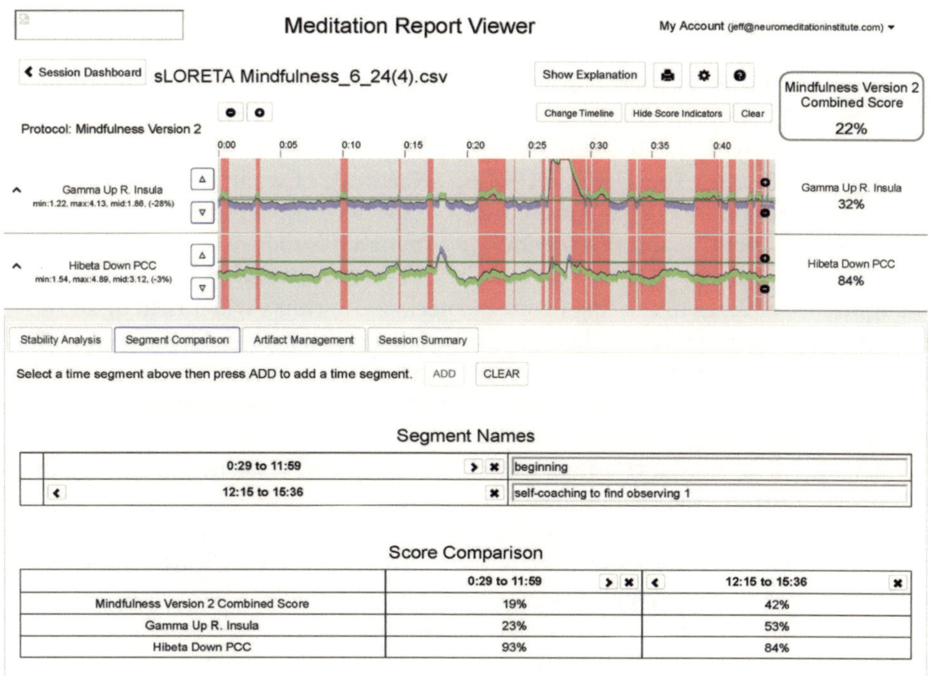


**Figure 7.**
*Analysis of session 4 in the NeuroMeditation report writer. Scores reflect the percent of time that EEG criterion was met for specified meditation segments.*

the EEG expectations such that she could use self-talk to help notice her present-moment experiences. This resulted in the most significant positive results to date. In fact, B.A. ended the session claiming that she felt "amazing." She described the meditative state as "feeling without trying to feel." She described it as an effortless awareness of her body in the present moment. **Figure 7** shows a comparison between the beginning of the session vs. the period of self-coaching. When this shift occurred in session, B.A.'s percent of combined success went from 19 to 42%.

#### *4.1.4.5 Sessions 5–8*

Once B.A. was able to experience the desired state and learn to do so in a way that felt safe without dissociating, she mastered it very quickly. During the next three sessions, she continued to demonstrate the ability to quickly find the desired meditative state and maintain it for increasing lengths of time. She also noted experiencing similar meditative moments through the day. B.A. reported feeling that she is fully "in her skin" and enjoying it. **Table 1** below is a comparison of sessions, 2, 4, 6, and 8. By examining the percent of success across each session, her progress is clear.

| **Score Comparison** | | | | |
|---|---|---|---|---|
| | **Session 2** | **Session 4** | **Session 6** | **Session 8** |
| Mindfulness Combined Score | 7% | 22% | 44% | 75% |
| Gamma Up *R. insula* | 15% | 32% | 45% | 83% |
| High Beta Down PCC | 61% | 84% | 100% | 86% |

**Table 1.**
*Comparison of sessions 2, 4, 6, and 8 in NeuroMeditation report writer.*

### *4.1.5 Pre-post symptom changes*

At the conclusion of each neuromeditation session, B.A. completed the Toronto Mindfulness Scale [40]. This is a 13-item self-report scale designed to assess state mindfulness with respect to meditation practice. Six items are summed to produce a total Curiosity score ($\alpha = 0.88$), reflecting an attitude of wanting to learn more about one's experiences (e.g., "I was curious to see what my mind was up to from moment to moment"), and seven items are summed to produce summed to produce a Decentering score ($\alpha = 0.82$), reflecting a shift from identifying personally with thoughts and feelings to relating to one's experience within a wider field of awareness (e.g., "I was aware of my thoughts and feelings without overidentifying with them"). **Figure 8** shows her scores for each session.

These results show a consistent increase in Decentering, which is the scale most relevant to decreasing stress and anxiety.

B.A. also completed a symptom questionnaire around sessions 4 and 8. **Figure 9** below demonstrates the change in symptoms from Pre-post.

The gray areas at the top of each bar represent the scores during the initial assessment, the colored areas represent the most recent scores. Clearly, there was a perceived decrease in symptoms, which were most notable in "easily distracted," "filtering," "hypervigilance," "reading comprehension," and "worry." Most of these improvements appear directly related to the goals and concerns identified in the intake process. It is unclear why there would be such improvements in reading comprehension. It is possible that improvements in attention resulted in improved reading comprehension. It is also possible that any brain changes

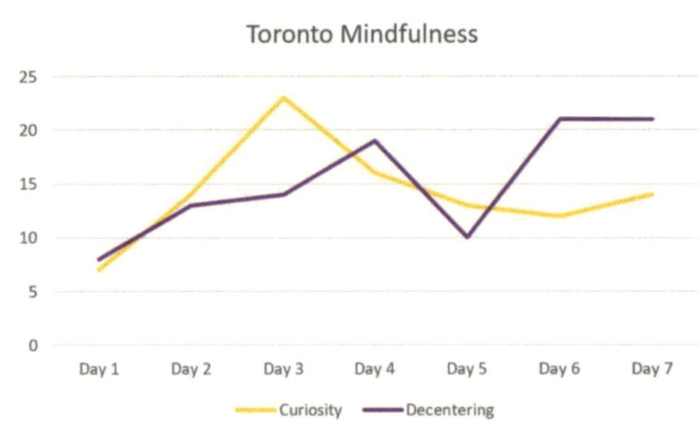


**Figure 8.**
*Toronto Mindfulness Scale scores (curiosity, decentering) for client B.A. across sessions 1–7.*

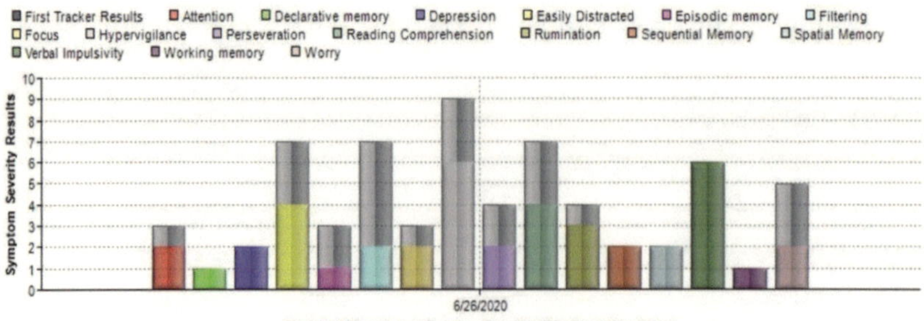


**Figure 9.**
*Session 4 vs. session 8 symptom questionnaire scores for client B.A.*

occurring as a result of the training had a more generalized impact on brain health and functioning, influencing concerns not directly related to the training itself.

## 5. Conclusions

While combining neurofeedback with meditation is not new, advances in our understanding of the neurological mechanisms of meditation have led to a more refined approach. Clinicians and researchers are now able to identify different meditation styles based not only on the way attention and intention are directed, but on brainwave patterns and brain regions involved. This has led to the ability to personalize the process, helping meditators choose a meditation style that is most likely to address their goals and needs. Indeed, researchers are now beginning to show that specific neuromeditation approaches can be used to improve cognitive functioning [41], and psychological concerns including anxiety, depression, and PTSD [5]. When this process is used in conjunction with meditation coaching, it is possible to use neuromeditation as a treatment modality that is individualized, and trauma informed. As such, neuromeditation promises to help define and refine meditation for the 21st century.

## Conflict of interest

The author is the Director of the neuromeditation Institute.

## Thanks

Special thanks to Ray Jackson for her assistance in manuscript preparation, B.A. for sharing her story, and Andrew Tang for development of the neuromeditation Report Writer.

## Author details

Jeff Tarrant
Neuromeditation Institute, Eugene, OR, United States

*Address all correspondence to: jeff@neuromeditationinstitute.com

## References

[1] Goyal M, Sing S, Sibinga E, Gould N, Rowland-Seymour A, Sharma R, et al. Meditation programs for psychological stress and well-being: A systematic review and meta-analysis. JAMA Internal Medicine. 2014;**174**(3):357-368. DOI: 10.1001/jamainternmed.201313018

[2] Vnuk V. Motivations for Meditating. Tucson, AZ: University of Arizona; 2016. Unpublished manuscript. These submitted to Honors College

[3] Brandmeyer T, Delorme A. Meditation and neurofeedback. Frontiers in Psychology. 2013;**4**:688. DOI: 10.3389/fpsyg.2013.00688

[4] Tarrant J. Meditation Interventions to Rewire the Brain: Integrating Neuroscience Strategies for ADHD, Anxiety, Depression and PTSD. Eau Claire, WI: PESI Publishing and Media; 2017

[5] Tarrant J. NeuroMeditation: An introduction and overview. In: Collura TF, Frederick JA, editors. Clinician's Companion to QEEG and Neurofeedback (Annotated and with an Introduction by J. Kiffer). New York: Taylor & Francis; 2017

[6] Evans J, Abarbanel A. Introduction to Quantitative EEG and Neurofeedback. Elsevier. Cambridge, MA: Academic Press; 1999

[7] Hammond CD. What is Neurofeedback: An update. Journal of Neurotherapy. 2011;**15**(4):305-336

[8] Travis F, Shear J. Focused attention, open monitoring, and automatic self-transcending: Categories to organize meditations from Vedic, Buddhis, and Chinese traditions. Consciousness and Cognition. 2010;**19**(4):1110-1118. DOI: 10.1016/j.concog.2010.01.007

[9] Anand B, Chhina G, Singh B. Some aspects of electroencephalographic studies in yogis. Electroencephalography and Clinical Neurophysiology. 1961;**13**(3):452-456. DOI: 10.1016/0013-4694(61)90015-3

[10] Kasamatsu A, Hirai T. An electroencephalographic study on the Zen meditation (Zazen). Psychiatry and Clinical Neurosciences. 1969;**20**(4): 315-336. DOI: 10.1111/j.1440-1819.1966.tb02646.x

[11] Crane R. Infinite potential: A neurofeedback pioneer looks back and ahead. In: Press THM, editor. Handbook of Neurofeedback: Dynamics and Clinical Applications. Binghamton, NY: The Haworth Press Inc; 2007. pp. 3-21

[12] Klimesch W. Alpha-band oscillations, attention, and controlled access to stored information. Trends in Cognitive Sciences. 2012;**16**(12):606-617

[13] Hardt J, Kamiya J. Anxiety change through electroencephalographic alpha feedback seen only in high anxiety subjects. Science. 1978;**201**(4350):79-81. DOI: 10.1126/science.663641

[14] Kamiya J. Operant control of the EEG alpha rhythm and some of its reported effects on consciousness. In: Altered States of Consciousness. New York, NY: Wiley; 1969. pp. 489-501

[15] Hall GB. Sagittal MRI Slice with Highlighting Indicating Location of the Posterior Cingulate Cortex. [MRI image]. 2011. Available from: https://en.wikipedia.org/wiki/Anterior_cingulate_cortex#/media/File:MRI_anterior_cingulate.png

[16] Andrews-Hanna JR. The brain's default network and its adaptive role in internal mentation. The Neuroscientist: A Review Journal Bringing Neurobiology, Neurology and Psychiatry. 2012;**18**(3):251-270. DOI: 10.1177/1073858411403316

[17] Smallwood J, Nind L, O'Connor R. When is your head at? An exploration of the factors associated with the temporal focus of the wandering mind. Consciousness and Cognition. 2009;**18**(1):118-125. DOI: 10.1016/j.concog.2008.11.004

[18] Garrison K, Santoyo J, Davis J, Thornhill T, Kerr C, Brewer J. Effortless awareness: Using real time neurofeedback to investigate correlates of posterior cingulate cortex activity in meditators' self-report. Frontiers in Human Neuroscience. 2013;**7**:440. DOI: 10.3389/fnhum.2013.00440

[19] Garrison K, Scheinost D, Worhunsky P, Elwafi H, Thornhill T, Thompson E, et al. Real-time fMRI links subjective experience with brain activity during focused attention. NeuroImage. 2013;**81**:110-118. DOI: 10.1016/j.neuroimage.2013.05.030

[20] Van Lutterveld R et al. Source-space EEG neurofeedback links subjective experience with brain activity during effortless awareness meditation. NeuroImage. 1 May 2017. 2016;**151**:117-127. DOI: 10.1016/j.neuroimage.2016.02.047

[21] Brandmeyer T, Delorme A. Closed-loop frontal midlineθ neurofeedback: A novel approach for training focused-attention meditation. Frontiers in Human Neuroscience. 2020;**14**:246. DOI: 10.3389/fnhum.2020.00246

[22] Aftanas L, Lotova N, Koshkarov V, Popov S. Non-linear dynamical coupling between different brain areas during evoked emotions: An EEG investigation. Biological Psychology. 1998;**48**(2):121-138. DOI: 10.1016/S0301-0511(98)00015-5

[23] Aftanas L, Varlamov A, Pavlov S. Affective picture processing: Event-related synchronization within individually defined human theta band is modulated by valence dimension. Neuroscience Letters. 2001;**303**(2):115-118. DOI: 10.1016/S0304-3940(01)01703-7

[24] Başar E, Schürmann M, Sakowitz O. The selectively distributed theta system: Functions. International Journal of Psychophysiology. 2001;**39**(2-3):197-212

[25] Dietl T, Dirlich G, Vogl L, Lechner C, Strian F. Orienting response and frontal midline theta activity: A somatosensory spectral perturbation study. Clinical Neurophysiology. 2009;**110**(7):1204-1209. DOI: 10.1016/S1388-2457(99)00057-7

[26] Aftanas L, Golocheikine S. Human anterior and frontal midline theta and lower alpha reflect emotionally positive state and internalized attention: High-resolution EEG investigation of meditation. Neuroscience Letters. 2001;**310**(1):57-60. DOI: 10.1016/S0304-3940(01)02094-8

[27] Brandmeyer T, Delorme A, Wahbeh H. The neuroscience of meditation: Classification, phenomenology, correlates, and mechanisms. Progress in Brain Research. 2019;**244**:1-29

[28] Tarrant J. FM Focus [Video]. Eugene, OR: The Neuromeditation Institute; 2020. Available from: https://www.youtube.com/watch?v=OxMdYj2Jq4Y&feature=youtu.be

[29] Fox K et al. Functional neuroanatomy of meditation: A review and meta-analysis of 78 functional neuroimaging investigations. Neuroscience and Biobehavioral Reviews. 2016;**65**:208-228

[30] Lutz A, Brefczynski-Lewis J, Johnstone T, Davidson R. Regulation of the neural circuitry of emotion by compassion meditation: Effects of meditative expertise. PLOS One. 2008;**3**(3):e1897. DOI: 10.1371/journal.pone.0001897

[31] Cahn B, Polich J. Meditation states and traits: EEG, ERP, and neuroimaging studies. Psychological Bulletin. 2006;**132**(2):180-211. DOI: 10.1037/0033-2909.132.2.180

[32] Lutz A, Dunne J, Davidson R. Meditation and neuroscience of consciousness: An introduction. In: Zelazo P, Moscovitch M, Thompson E, editors. The Cambridge Handbook of Consciousness. New York: Cambridge University Press; 2006

[33] Mindful. How the Brain Changes When You Meditate: By Charting New Pathways in the Brain, Mindfulness Can Change the Banter Inside Our Heads from Chaotic to Calm [Internet]. 2015. Available from: https://www.mindful.org/how-the-brain-changes-when-you-meditate/ [Accessed: 02 July 2020]

[34] Yogi M. Celebrating Perfection in Education: Maharshi Vedic. 2nd ed. Noida, India: Maharshi Vedic University Press; 1997

[35] Tarrant J. NeuroMeditation Styles Inventory [Internet]. 2017. Available from: https://www.neuromeditationinstitute.com/whats-your-style [Accessed: 25 May 2020]

[36] Soutar R. New Mind Cognitive Emotional Checklist (CEC) Symptom Checklist [Internet]. 2015. Available from: https://www.newmindmaps.com/UI/index.aspx?ReturnUrl=%2f [Accessed: 15 June 2020]

[37] Keizer A. qEEG-Pro Report Service [Internet]. 2013. Available from: https://qeeg.pro/ [Accessed: 07 June 2020]

[38] Craig A. How do you feel—Now? The anterior insula and human awareness. Nature Reviews Neuroscience. 2009;**10**(1):59-70. DOI: 10.1038/nrn2555

[39] Fleming S, Dolan R. The neural basis of metacognitive ability. Philosophical Transactions of the Royal Society, B: Biological Sciences. 2012;**367**(1594):1338-1349

[40] Lau M, Bishop S, Segal Z, Buis T, Anderson N, Carlson L, et al. The Toronto mindfulness scale: Development and validation. Journal of Clinical Psychology. 2006;**62**(12):1445-1467

[41] Thibault RT, Lifshitz M, Birbaumer N, Neurofeedback RA. Self-regulation, and brain imaging: Clinical science and fad in the service of mental disorders. Psychotherapy and Psychosomatics. 2015;**84**(4):193-207

Section 4

# Psychophysiological Psychotherapy

Chapter 4

# Biofeedback in Clinical Psychology: Modalities and Perspectives

*Valeska Kouzak, Aloysio Campos da Paz Neto and Ivo Donner*

## Abstract

Biofeedback is a technique of self-regulation applied by health professionals in order to reshape a series of physiological information based in health parameters diminishing psychopathological symptoms and improving cognitive performance. The biofeedback technique is widely recognized in many countries, leaving no doubt about its effectiveness and applicability. In clinical psychology, biofeedback has been applied effectively to psychophysiological conditions such as anxiety, depression and ADHD. This chapter has the aim to elucidate the techniques applied to clinical settings, where psychophysiological conditions are more prone to be treated with biofeedback. Moreover, this chapter also evaluates the advances of the technique and possible future directions.

**Keywords:** neurofeedback, HRV, GSR, anxiety, depression, ADHD

## 1. Introduction

Biofeedback has begun in the early 60's as a result of the convergence of diverse and relatively recently established fields, such as bio-engineering, physiology and psychology. Firstly, biofeedback emerged as a technique of stress control when studies relating psychological condition and physical disease emerged more vigorously. Afterwards, with the development of self-regulation techniques for managing physical and emotional burden, biofeedback became more sophisticated in its applications and methods.

Technological advances in instrumentation led to more objective measurements of physiological signals, and the measurement + feedback of a subject's performance has been established to improve capacity to control one's own signals [1]. This way, the organism's immediate response has been used as a live indication of self-regulation during training.

Diverse instruments have been applied to self-regulation, such as skin conductance response, body temperature, cardiac frequency, respiratory frequency, electroencephalography (EEG) [2], and recently, functional magnetic resonance imaging. The combination of these instruments with theory grounded biofeedback's basic principle, which is to provide awareness about one's physical condition - creating a self-awareness loop - and through the technology, allow psycho-physiological self-control. Hence, biofeedback provides an organism with a self-referential information feedback to be adjusted according to personal needs [3].

In this sense, it is important to make a distinction between treatment and training: treatment is a process in which an individual expects an external agent to provide the change for release or suppression of symptoms; on the other hand, training asks the individual for their voluntary participation to the same outcome. In this chapter biofeedback is considered a training protocol [4] as it requires an active participation of the patient in the process.

This technique is characterized by a multidisciplinary and heterogeneous application, and in psychology, biofeedback is applied to relive symptoms of many psycho-physiological diseases, such as ADHD, anxiety, depression, among others [5–9]. Besides that, nowadays biofeedback is also applied as technique for stress reduction in non-clinical population - mainly those that have in their routine highly demanding situations of emotional control - such as athletes, police-officers, the military, executives, students, musicians, for increasing their performance [10–12].

Moreover, there are studies that seek to evaluate the effects of neurofeedback training in cognitive and neuro-psychological abilities. Neurofeedback is a form of biofeedback that regulates cortical activity using EEG feedback of brain waves. Results so far as 2020 have demonstrated an increase of attention [13], working memory [14] and executive function performance [15] in cognitive and neuro-psychological abilities.

As a multidisciplinary area, biofeedback education and training for professionals are diverse, existing in the forms of workshops and as university courses. Hence, in order to regulate the quality of practitioners, societies have been created, such as the Biofeedback Certification International Alliance (BCIA) - one of the most widely known and recognized - to evaluate and regulate the knowledge and practice of professionals, and provide support for them and their patients.

In summary, the current chapter has the objective to briefly explain through examples of biofeedback techniques how biofeedback has been used in psychology to ameliorate psychophysiological symptoms of diverse pathologies, as well as how it has been used as an instrument to reduce stress and enhance performance.

## 2. Basic principles of psychophysiology

The autonomic nervous system (ANS) is a regulatory part of our nervous system that works to ensure homeostasis - the maintenance of our body's balance. Therefore, vegetative activity is regulated by the ANS. Body temperature, respiration rate, heart rate, digestion, secretions and muscles are regulated by the ANS's activity unconsciously in order to keep the body alive [16].

The ANS is controlled by the central nervous system (CNS) from sub-cortical areas (hypothalamus, brain stem, spinal cord and others) but some cortical areas also control its activity. Hence, ANS activity is not completely involuntary [17].

The ANS can be divided in two pathways:

a. Sympathetic Pathway: responsible for increasing body activity. To that end, the Sympathetic Pathway is also responsible for our fight-or-flight response, such as increased heart rate frequency, sweat and respiration, or promoting muscular contraction.

b. Parasympathetic Pathway: responsible for decreasing body activity. To that end, the Parasympathetic Pathway promotes relaxation of our body, such as decreasing heart rate frequency and respiration or allowing muscular relaxation.

Sympathetic and parasympathetic pathways work together to coordinate organism function, activating and deactivating over time to maintain body balance.

The balance between sympathetic and parasympathetic pathways may be briefly explained by the relation between respiration sinus arrhythmia (RSA), that is, the influence of sinuatrial node of the heart by inhaling and exhaling movement in heart rate frequency. Therefore, when individuals inhale their heart rate frequency increase and when they exhale their heart rate decrease, leading to a continuous sympathetic and parasympathetic activity, activating and deactivating our body [1]. However, when changes in our homeostasis happens, more sympathetic activity is required, changing our respiration for a fast paced and higher inhalation, consequently our heart rate frequency increase, our muscles contracts, our body temperature decrease, among others physiological changes. As soon as individuals recognize and classify the stressor being a threat or not, the balance can be restored [17].

Biofeedback works by regulating the ANS response to the environment, providing balance between sympathetic and parasympathetic responses. Consequently, another important concept associated to biofeedback is stress [18]. Stress can be defined as a body response to external or internal changes affecting homeostasis. The cause of stress is defined as the stressor. Usually, stress is associated to a negative stressor. However, stressors can be all things that affect organism balance, there existing positive stressors as well (i.e: marriage, born of a child, graduation presentation). Generally, a stressor can be physical (i.e: broken bone), metabolic (i.e: diabetes), biological (i.e: Virus), mental (i.e: depression), social (i.e: divorce) and cultural (i.e: civil war) [18].

Biofeedback can create awareness of this relationship between stressors and body response, helping individuals to regulate their conditions and foster well-being.

## 3. Modalities and perspectives

In this section we intend to provide the most common modalities of biofeedback, equipment and the perspective of how it has been applied to manage the psychophysiological conditions. The modalities are presented in following order:

1. Galvanic Skin Response (GSR);

2. Temperature Biofeedback;

3. Respiratory biofeedback;

4. Surface Electromyography Biofeedback (sEMG);

5. Heart Rate Variability Biofeedback;

6. Electroencephalography (EEG) Biofeedback or Neurofeedback;

7. Functional Magnetic Resonance Neurofeedback (rt-FMRI);

8. Functional near-infrared spectroscopy (fNIRS) Neurofeedback.

### 3.1 Galvanic skin response (GSR) biofeedback

One of the first equipment used to demonstrate the correlation between mood alterations and physiological changes was the Galvanic Skin Response (GSR). The physiological principle that ground this equipment is the direct correlation between stress and autonomic response of sweat by the organism. That is, the higher is the stress, higher is the sweat of the organism, and vice-versa.

Since sweat is a compound of mineral salt and water, and an excellent electrical conductor, its secretion enhances galvanic skin conductance, which can be measured by two ultra-sensitive sensors to electric current [19].

GSR works based on the following arrangement: the device produces a non-invasive electrical current of a minimal value, imperceptible to the organism, that varies according to skin resistance change with the sweat excretion amount [1]. This electrical current might be converted in sounds and/or images that modify according to amount changes in skin sweat. In other words, in a sound feedback, the device might emit different sound frequencies according the skin conductance, such as:

a. A high-pitched sound is emitted when there are an increase of the sweat amount;

b. A bass sound is emitted when there is a decrease of the sweat amount.

Consequently, the bass sound would correspond to lower levels of anxiety and pitched sound would correspond to higher levels of anxiety. Hence, the sounds allow the representation of subject's physiological response in relation their own levels of anxiety and stress [20].

The GSR biofeedback training has been applied to manage stress in anxious patients [21], medication resistant epilepsy patient [22] and athletes [23]. In clinical setting, this modality is usually applied with other biofeedback technique, such as temperature biofeedback or respiratory biofeedback.

### 3.2 Temperature biofeedback

This biofeedback technique uses body temperature as a parameter, with measurements in the majority of cases from the extremities of superior members - usually fingers tips or hands. Body temperature is also regulated by autonomic response, and a reduction of blood flow in these areas is an indication of high stress level whereas passivity is related to higher temperature. Therefore the relationship between body temperature and stress is inverse: the higher is the stress, the lower is the body temperature [24].

In other words, when blood vessels contract due to stress increase, blood flow and therefore temperature diminishes. On the other hand, when the individual is relaxed, there is a vasodilation and an increase of blood flow and therefore of temperature.

The temperature biofeedback technique is used for relaxation, being a quickly responsive indicator, very sensitive to stress variation [21]. This biofeedback training can also be arranged to provide sound feedback. For instance, if the individual needs to manage stress, a higher pitch sound is played for low temperatures and a bass sound for higher temperatures.

In summary, the skin temperature biofeedback consists in regulating the body temperature in order to change the autonomic system. It is rarely used alone in clinical setting, which most of the time is associated with other technique such as respiration biofeedback to reduce stress symptoms in anxious patient.

### 3.3 Respiratory biofeedback

Another technique of biofeedback is the respiratory. It has the aim to improve ventilation capacity, facilitate carbon dioxide elimination, and develop relaxation skills to manage stress and other symptoms [25]. This technique is fundamental because it allows an indirect modulation of the autonomic nervous system, mainly because inspiration activates the sympathetic pathway and mobilizes the organism for action; on the other hand, expiration stimulates the parasympathetic pathway, that is involved in homeostatic process (psychophysiological balance) and general body relaxation [1, 16, 26].

Respiratory biofeedback training can be done using belt-coupled sensors over the thorax circumference and/or abdominal circumference to register data and repass it to an encoder that will present information in a comprehensive by way of a computer, through bars, graphic, counters, percentage, or even a sound.

To that way, the professional and patient will be able to analyze the respiratory frequency (number of respiratory cycles per minute) and the respiratory amplitude (volume of air change) facilitating the regulation of anxiety symptoms for instance [1, 25].

In addition, it is important to observe that the thorax belts or abdominal belts contribute to the muscular contraction perception, as lungs are dependent on other muscles to breathe. Therefore, respiratory biofeedback technology is able to analyze abnormal respiratory cycles, such as:

a. Tachypnea (fast respiration, superficial, with high frequency) and;

b. Hyperventilation (deep, fast respiration, with increase of respiration frequency and volume).

In both cases, the fast respiration results in unbalanced entrance of carbon dioxide and oxygen in the blood, and in diminishing carbon dioxide, the blood becomes alkaline. That is, the balance of calcium and potassium in cellular membranes is modified and lead to muscular and nervous dysfunction, such as mental fatigue, migraine, dizziness, syncope among others [27, 28].

Respiratory biofeedback is a technique that can be associated with other biofeedback techniques to improve treatment effects on humor disorder, anxiety, cardiovascular and neurological disease, cognitive and physical performance.

### 3.4 Surface electromyography biofeedback (sEMG)

Surface electromyography measures electrical activity of muscles by sensors placed over the skin, where the activity should be measured. Muscular tension signs captured by electrodes are translated in sounds and images, similar to GSR biofeedback. Therefore, depending of the training, the following arrangement can be proposed:

a. To problems related to muscular tension, for instance temporomandibular joint muscle tension, the feedback is related to facial muscle relaxation. The feedback might be a pitched sound for muscular tension, and when the muscle is relaxed, it might be a bass sound.

b. To problems related to flaccid muscle, for instance neurological lesion that induces to muscle tone loss, the feedback is related to muscle contraction for strengthening tone. That is, when the muscle is contracted the feedback might be a bass sound, on the other hand when the muscle is relaxed, than the feedback might be a pitched sound.

Consequently, the sEMG biofeedback is used not only for general relaxation training, but also to treat facial paralysis or movement rehabilitation in cases of brain damage.

### 3.5 Heart rate variability biofeedback

Heart rate variability describes oscillations between heart rate over consecutive intervals, that is, a R-R wave cycle, controlled by autonomic nervous system (ANS) over the sinusoidal node. The cardiac frequency is regulated by sympathetic and parasympathetic pathways. In other words, the increase of cardiac frequency is related to a higher action of sympathetic pathway and the decrease of cardiac frequency is related to a higher action of parasympathetic pathway. Heart rate variability demonstrates heart health, as it is not expected that the heart keeps the same frequency, but instead might be able to adjust to adverse condition that is presented to the individual [29].

Heart rate variability training is accomplished when the cardiac rhythm enters in consonance or synchronicity with the respiratory rhythm, increasing the amplitude of heart rate oscillation [29]. This variability is maximized by biofeedback training when six respiratory cycles per minute is achieved, which leads to a sinusoidal wave similar to sinus arrhythmia [30].

The training is accomplished by observation of cardiac frequency waves and respiration and as the waves became consonants, that is, overlapping or "in phase", sound or visual feedback is offered as behavioral reinforcement.

Heart rate variability biofeedback is related to decreased symptoms of many clinical conditions such as anxiety, depression, post-traumatic stress disorder, insomnia, alcoholism and addiction, mostly because it provides a feeling of well-being [2, 31].

Besides that, in studies of increasing performance, the sensation of well-being induced by the heart rate training demonstrated to be positive for professional musicians and dancers [32].

### 3.6 Electroencephalography (EEG) biofeedback or neurofeedback

In the beginning of the 20th century, Hans Berger, a German Neuropsychiatrist, investigated cerebral electrical activity in dogs and other species. From 1920 onward, Berger's studies evolved, and through modifications in electrodes and decoders, it was possible to finally use an electroencephalogram in humans in 1924, when Berger used it for the first time despite technical limitations. In 1930, Berger published his data about brain wave frequencies registered in his research, in which he named brain waves in accordance with Greek letters "alpha" and "beta" [33, 34].

During many years such knowledge of brain waves had mainly been for laboratory use. Until the evolution of equipment, applications were unknown.

At the 60's, Barry Sterman investigated brain waves patterns in training cats via a behavioral technique called operand conditioning, in which the animal response of pressing a bar is increased in frequency according to positive reinforcement, which could be food for instance. Sterman realized that when cats are prepared to press the bar, there was an increase of cortical activity in the sensorimotor area to a pattern of 12–14 Hz, together with a reduction of motor activity followed by muscular relaxation. This pattern of cortical activity is denominated the Sensorimotor Rhythm (SMR) [35, 36].

Brain wave frequencies are today classified in different groups defined by frequency bands and named as:

- Delta (1-4 Hz)
- Theta (4–7 Hz)
- Alpha (8-13 Hz)
- Beta (13-)
- Gamma (21-45 Hz)
- SMR (12–15 Hz) in sensorimotor area [36].

Moreover, each frequency is associated to specific cognitive, emotional and behavioral states. Delta waves are associated with sleep; Theta and Alpha waves are associated with working memory, attention and creativity; Beta waves are associated with intensive cognitive activity and memory; Gamma waves are associated with cognitive integration [8, 25].

The EEG Biofeedback training is also called neurofeedback, and can be defined as the operant conditioning of brain waves activity by EEG. That is, neurofeedback is the behavioral, cognitive or emotional training during electrophysiologic activity evaluation. This measure provides a feedback to the subject of their own performance, and, as consequence, provides awareness of their own state and of how to efficiently achieve an objective. [37].

The neurofeedback arrangement is based on electrodes that are placed in the scalp according to international 10/20 system, a converter decomposing the electrical activity from cortical neurons in frequency bands as stated above, and a software transforming raw information into computer actions (movie, games or music) for the training goal. For instance, if an individual has the aim to train attention, a protocol might be applied in the central cortex - therefore an electrode is placed at the Cz area - and software is programmed to only provide feedback when increased activity is detected in a band frequency related to increased attention (i.e. SMR) [38, 39].

Neurofeedback training became very popular and many researchers and clinicians had the opportunity to add and validate data and results. Neurofeedback is nowadays tested as a tool to treat ADHD and seizure with specificity and efficacy. Therefore, Neurofeedback studies to treat ADHD and seizure have demonstrated statistical significance in comparative studies such as patient-placebo, neurofeedback-medication, and neurofeedback-other treatments in more than two studies [40]. Neurofeedback is also becoming frequently applied in clinics to treat other pathologies with psychophysiological basis, such as, anxiety, depression and insomnia [9, 41, 42].

Therefore, the existing references are sufficient to reinforce its clinical use as tool and its increment in research, mainly in the USA, Canada and some European countries, such as England, Spain, Italy, Sweden and Austria [6, 36].

### 3.7 Functional magnetic resonance neurofeedback (rt-FMRI)

Another form of neurofeedback is by Functional Magnetic Resonance Image (FMRI), that offers data of cerebral activity from hemodynamic signal - Blood Oxygenation Level-Dependent (BOLD). Neurofeedback by FMRI is called real time Functional Magnetic Resonance Image (rt-FMRI) and has the advantage of providing training in deeper and specific regions of the brain and needing fewer sessions. Although it is a new technique, it results have been significant [43] in cognitive training.

The principle behind FMRI neurofeedback is that magnetic resonance captures changes in blood flow during brain activity and feedback on this change can be offered to the individual. In neural activation, metabolic demand increases, and consequently, oxygenated blood increases to regulate the de-oxygenated arterial blood. The increase of oxi-deoxyhemoglobin also increases the resonance signal around the activated nervous tissue, so the target area is indicated and a feedback is provided about this event.

FMRI neurofeedback is highly spatially precise. On the other hand, it lacks temporal precision, and therefore to capture an image it is necessary to wait for the neuronal dynamics to happen, which takes a few minutes. This makes the FMRI feedback slower than EEG. Therefore, the protocols used in FMRI neurofeedback must contemplate this temporal delay [44]. It is a new technique that has proposed protocols so far to improve performance, as it requires a very stressful environment to accomplish the training, which is the magnetic resonance machine, usually in hospitals.

### 3.8 Functional near-infrared spectroscopy (fNIRS) neurofeedback

The register by functional near-infrared spectroscopy (fNIRS) is similar to the magnetic resonance in that brain activity is captured from a BOLD signal. However, it differs in that the hemodynamic changes are registered by an infrared light spectrum only able to capture activity from layers closer to scalp [45].

The arrangement for this technique consists of a belt with fNIRS sensor placed over the scalp, and a computer feedback provided when individuals achieve the training goal.

Moreover, neurofeedback training by fNIRS is also recent and has been applied to treat depression [46] and improve cognitive performance [47]. It has the advantage of being more accessible, as it involves simpler equipment and less medical environment compared to rt-fMRI.

## 4. Performance optimization biofeedback

Nowadays, as the biofeedback class developed and became used in diverse applications, research about its use in non-clinical populations [48], such as with athletes, musicians, dancers and executives, have the aim to increase cognitive and physical performance [32]. Becoming an efficient tool to sports psychologists that face the need to help athletes to manage stress in order to better perform [49].

There are also studies that search for neurofeedback training in cognitive and neuropsychological skills. The results, as of 2020, have been demonstrating an increase in attention [13], working memory [14] and executive function [15].

In addition, there are studies in aging populations to provide cognitive reserve as their higher life expectancy increases the need of an active aging population [50].

## 5. Clinical application in psychology

According to the biofeedback modalities descriptions stated in the third topic of the present chapter, it is possible to enumerate the clinical application of biofeedback in psychology.

Firstly, to manage stress in anxious patient, which has been the most widespread application of biofeedback since the publication of Hardt and Kamiya study in 1978 [5] by managing alpha waves in neurofeedback training, as well as, in Leher protocol training of HRV [51], in which it is achieved when breath and heart rate frequency are consonant.

Secondly, to manage hyperactivity of children ADHD children in SMR protocol developed by Lubar and Lubar [6] and applied consequently to manage attention [9] and associate to metacognitive strategies [7].

Thirdly, with the acknowledge that depression is associate with activation differences between hemispheres, therefore Hammond proposed a training protocol to balance this interhemispheric difference applied frontally in the cortex [41].

Hence, those protocols are examples of successful training that have been applied by many biofeedback clinician as a psychological tool to manage symptoms and decrease suffering. Moreover, it is important to highlight how important is to associate psychotherapeutic approach to training, as understand psychological and psychotherapeutic aspects help clinicians to improve rapport and coping [25].

## 6. Conclusion

The connection between body and mind is well known in psychology and has been a subject of discussion and academic work. Many mental disorders have their origin in the central nervous system, and on the other hand the environment also has a great impact in our body. Therefore, a biopsychosocial perspective is needed in health professionals, to fully attend individuals that search for clinical health in psychology. To that extend, biofeedback is a tool that highlights the value of body regulation to provide well-being.

The chapter aimed to demonstrated the diverse biofeedback techniques that provide self-awareness and help patients to cope with, and manage, symptoms.

Most of the techniques are applied in regulation of sympathetic symptoms activated by stress, thus being relevant to coping with a highly demanding contemporary life. Self-regulation of stress is connected with the amelioration of anxiety symptoms and attention deficits, while providing better emotional and cognitive awareness.

## Author details

Valeska Kouzak[1*], Aloysio Campos da Paz Neto[1] and Ivo Donner[2]

1 University of Brasília, Brasília, Brazil

2 Donner Health Center, Brasília, Brazil

*Address all correspondence to: valeskakcp@gmail.com

## References

[1] Schwartz, M; Andrasik (2003), F. *Biofeedback: A Parctioner's guide*, 3o. Ed. Nova Iorque, EUA: Guildford Press.

[2] Lantyer, A.; Viana, M.; e Padovani, R. (2013) Biofeedback no tratamento dos transtornos relacionados ao estresse e á ansiedade: uma revisão critica, *Psico-USP*, Bragança Paulista, *v.18 (1).*

[3] Wheat, A.L, Larkin, K. (2010) Biofeedback of Heart Rate Variability and Related Physiology: A critical review, *Applied Psychophysiology Biofeedback*.

[4] Jaeggi, S.M, Bushkuhel, M., Jonides, J., Shah, P., (2011) Short and Long benefits of cognitive training. *PNAS, v. 25 (108), 10081-10086*

[5] Hardt, J. V., Kamiya, J. Anxiety change through electroencephalographic alphafeedback seen only in high anxiety subjects. Science. vol. 201, p. 79-81, 1978.

[6] Lubar, J.O; Lubar, J.F. (1984) Electroencephalographic Biofeedback of SMR and beta for treatment of attention deficit disorders in a clinical setting, *Applied psychophysiology and Biofeedback, v. 9 (1), 1-23*

[7] Thompson, M; Thompson, L. (1998); Neurofeedback combine with training in metacognitive strategies: Effectiveness in student with ADD. *Applied Psychophysiology and Biofeedback,* v. *23 (4).*

[8] Hammond, D. C. (2006). What is Neurofeedback? *Journal of Neurotherapy*. Haworth Press, *v. 10 (4).*

[9] Monastra V. (2005) Electroencephalographic Biofeedback in the treatment of attention-deficit/ Hyperactivity disorder. *Applied Psychophysiology and Biofeedback*, v 30 (2).

[10] Vernon, D., Egner, T., Cooper, N., Compton, T., Neilands, C., Sheri, A., Gruzelier,, J. The effects of training distinct neurofeedback protocols aspects of cognitive performance. International Journal of Psychophysiology, v. 47, 2003

[11] Gruzelier, J., Egner, T., Vernon, D. Validating the efficacy of neurofeedback for optimizing performance, 2006. http://research.gold.ac.uk/500/1/PSY_Gruzelier_2006a.pdf

[12] Gruzelier J. (2008); A theory of alpha/theta Neurofeedback, creative performance enhancement, long distance functional connectivity and psychological integration, Research Report, *Cognitive process*.

[13] Klimesch, W., Doppelmayr, M., Russengger, H., Pachinger, T., Schwaiger, J. Induced alpha band power changes in human EEG and attention. Neuroscience Letters, v. 244, n. 2, p. 73-76, 1998

[14] Klingberg, T. (2010) Training and plasticity of working memory. Trends Cogn Sci. 2010 Jul;14(7):317-24. doi: 10.1016/j.tics.2010.05.002.

[15] Enriquez-Geppert, S., Huster R.J., Herrman, J. (2013); Boosting Brain Function: improving executive function with behavioral training, neurostimulation and Neurofeedback; *International Journal of psychophysiology*.

[16] Carlson, N. R. (2002) *Fisiologia do comportamento*. 7ª. ed. São Paulo, SP: Manole. Brazil

[17] Lent, Roberto (2010) *Cem bilhões de neurônios? Conceitos fundamentais de neurociências*. 2nd ed. São Paulo, SP: Athena. Brazil

[18] Slavikova M., Sekaninova, N., Olexova L., Visnovicova, Z., Tonhajzerova, I (2020) Biofeedback- a

promising non pharmacological tool of stress-related disorders. *Acta Medica Martiniana*. Vol 20(1)

[19] Sharma, M., Kacker, S., Sharma, M. (2016) A brief introduction and review on galvanic skin response. *International journal of medical research professionals*. 2(6): 13-17.

[20] Thompson, M. Thompson, L. (2003) *The Neurofeedback Book-An introduction to basic concepts in applied psychophysiology*. Colorado, EUA: The Applied Psychophysiology and Biofeedback Publisher.

[21] Donner, I. O. (2011) Biofeedback *In:* Rangé B. et al. *Psicoterapias Cognitivos-Comportamentais: um dialogo com a psiquiatria*, 222-237, 2o. Porto Alegre, RS: ArtMed.

[22] Nagai, Y., Jones, C.I, Arjune, S. (2019) Galvanic Skin Response (GSR)/ Electrodermal/Skin Conductance Biofeedback on Epilepsy: A systematic review and meta-analysis. Frontiers in Neurology. V. 10. doi: 10.3389/fneur.2019.00377

[23] Pusenjak, N., Grad, A., Tusak, M., Leskovesek, M., Schwarzlin, R. (2015) Can biofeedback training of psychophysiological response enhance athletes' sport performance? A practitioner's perspective. *The physician and sports medicine*. doi: 10.1080/00913847.2015.1069169

[24] Shaffer, F., Combatalade, D., Peper, E. (2016) A guide to cleaner skin temperature recordings and more versatile use of your thermistor. *Biofeedback* v.44 (3): 168-176

[25] Demos, J. N. (2004). *Getting Started with Neurofeedback*. Nova York, NJ: W. W. Norton & Company.

[26] Gazzaniga, M., Ivry, R., Mangun, G. (2019) *Cognitive Neuroscience- The biology of the mind.* 5th ed, New York, NJ: W. W. Norton & Company.

[27] West, B. Jonh. (1996) *Fisiologia Respiratória Moderna- principios básicos*. 9th ed, Porto Alegre, RS: Artmed.

[28] Costa, E. L. V., Júnior, L. P. (2015). *Pneumologia Ventilação Mecânica: princípios e aplicação* 1°. Ed. São Paulo, SP: Atheneu.

[29] Lehrer, P. e Gavirtz. (2014) Heart Rate Variability Biofeedback: how and why does it works?. *Frontiers in Psychology, v 05.*

[30] Leher, P., Vaschilo, E., Shou-En, L., Eckberg, D., Vaschilo, B., Scardella, A., Habib, R. (2005) Heart rate variability-Effects of age on heart rate variability, baroreflex gain and asthma. *Chest Journal*, v. 129 (2).

[31] Gomes, J.S., Coghi, M.F., Coghi P.F. (2014) Biofeedback Cardiovascular e suas aplicações: revisão da literatura. *Avances en Psicologia Latinoamericana, v. 32 (2), 199-216.*

[32] Gruzelier, J. (2013) EEG-Neurofeedback for optimising performance. I: A review of cognitive and affective outcome in healthy participants; *Neuroscience and Biobehavioral reviews*.

[33] Ferreira, L.S. (2010). Aspectos históricos do EEG. In: Ferreira, L.S., Oliveira, P.A. L., Bonavides, A.S. *Manual do técnico em EEG*. Rio de Janeiro, RJ: Revinter.

[34] Cripps, A. A. S., Domingos C.J. & Paola, L. (2013). Eletroencefalografia. *In*: Brasil-Neto, J. P. & Takayanagui, O.M. *Tratado de Neurologia da Academia Brasileira de Neurologia*. Rio de Janeiro, RJ: Elsevier.

[35] Gruzelier,, J. EEG-neurofeedback for optimising performance. I: A review

of cognitive and affective outcome in healthy participants. Neuroscience and Biobehavioral reviews. 2013.

[36] Sterman, B.S. (2010). Biofeedback in the treatment of epilepsy. *Cleveland Clinic Journal of Medicine. v. 77(3).*

[37] Strehl, U. What learning theories can teach us in designing neurofeedback treatments. Frontiers in human neuroscience. v. 8, n. 894, 2014

[38] Bazanova, O. M., and Aftanas, L. I. (2010). Individual EEG a activity analysis for enhancement neurofeedback efficiency: two case studies. *J. Neurother.* 14, 244-253. doi: 10.1080/10874208.2010.501517

[39] Sitaram R., Ros T., Stoeckel L., Haller S., Scharnowski F., Lewis-Peacock J., et al., (2017) Closed-loop brain training: The science of neurofeedback. Nat Rev Neurosci. 18, 86-100. doi: 10.1038/nrn.2016.164.

[40] Arns, M. Ridder, S., Strehl, U., Breteler, M., Coenen (2009) Efficacy of Neurofeedback Treatment in ADHD: The effects of inattention, impulsivity and hyperactivity: a Meta-analysis. *Clinical EEG and Neurosciences. v. 40 (3),* 180-189.

[41] Hammond, D. (2005) Neurofeedback Treatment of Depression and Anxiety. *Journal of Adult Development, Vol. 12, Nos. 2/3,* DOI: 10.1007/s10804-005-7029-5

[42] Coortos, A., Valck, E., Arns, M., Breteler, M., Cluydits, R. (2010) An exploratory study on the effects of tele-neurofeedback and rele-biofeedback on objective and subjective Sleep in patients with primary insomnia. Appl Psychophysiol Biofeedback 35:125-134 DOI 10.1007/s10484-009-9116-z

[43] Soares, J.M., Magalhães, R., Moreira, P.S., Sousa, A., Ganz, E., Sampaio, A., Alves, V., Marques, P., Sousa, N. A hitchhiker's guide to functional magnetic resonance. Frontiers in Neuroscience. v.10. 2016

[44] Sherwood, M. S., Kane, J. H., Weisend, M. P., Parker, J. G. Enhanced control of dorsolateral prefrontal cortex neurophysiology with real-time functional magnetic resonance imaging (rt-fMRI) neurofeedback training and working memory practice. Neuroimage. v. 124, p. 214-223, 2016

[45] Kober S.E., Witte M., Stangl M., Väljamäe A., Neuper C., Wood G. (2014) Shutting down sensorimotor interference unblocks the networks for stimulus processing: An SMR neurofeedback training study. Clin Neurophysiol. 126, 82-95. doi: 10.1016/j.clinph.2014.03.031.

[46] Trambaioli, L.R., Kohl, S.H., Linden, D., Mehler, D. (2020) Neurofeedback training in major depressive disorder: a systematic review of clinical efficacy, study quality and reporting practices. https://osf.io/y48zw/

[47] Li, K., Jiang, Y., Gong, Y., Zhao, W., Zhao, Z., Liu, X., Kendrick, K., Zhu, C., Becker, B. (2019) Functional Near-Infrared Spectroscopy (fNIRS) informed neurofeedback: regional specific modulation of lateral orbitofrontal activation and cognitive flexibility. https://www.biorxiv.org/content/biorxiv/early/2019/01/04/511824.full.pdf

[48] Edmonds, W. A., & Tenenbaum, G. (Eds.). (2011). Case studies in applied psychophysiology: Neurofeedback and biofeedback treatments for advances in human performance. John Wiley & Sons.

[49] Landers, D. M. (1985). Psychophysiological assessment and biofeedback. In Biofeedback and sports science (pp. 63-105). Springer, Boston, MA.

[50] Campos da Paz, V.K., Garcia, A., Campos da Paz Neto, A., Tomaz, C. SMR neurofeedback facilitates working memory performance in healthy older adults: a behavior and EEG study. Frontiers in Behavioral Neuroscience. volume 12. 2018. doi: 10.3389/fnbeh.2018.00321

[51] Lehrer, P. (2003). Applied psychophysiology: Beyond the boundaries of biofeedback (mending a wall, a brief history of our field, and applications to control of the muscles and cardiorespiratory systems). Applied Psychophysiology and biofeedback, 28(4), 291-304.

Section 5

# Physiotherapy

Chapter 5

# Neurofeedback for Chronic Pain

*Kajal Patel, Manoj Sivan, James Henshaw and Anthony Jones*

## Abstract

Neurofeedback is a novel neuromodulatory therapy where individuals are given real-time feedback regarding their brain neurophysiological signals in order to increase volitional control over their brain activity. Such biofeedback platform can be used to increase an individual's resilience to pain as chronic pain has been associated with abnormal central processing of ascending pain signals. Neurofeedback can be provided based on electroencephalogram (EEG) or functional magnetic resonance imaging (fMRI) recordings of an individual. Target brain rhythms commonly used in EEG neurofeedback for chronic pain include theta, alpha, beta and sensorimotor rhythms. Such training has not only been shown to improve pain in a variety of pain conditions such as central neuropathic pain, fibromyalgia, traumatic brain injury and chemotherapy induced peripheral neuropathy, but has also been shown to improve pain associated symptoms such as sleep, fatigue, depression and anxiety. Adverse events associated with neurofeedback training are often self-limited and resolve with decreased frequency of training. Provision of such training has also been explored in the home setting whereby individuals have been encouraged to practice this as and when required with promising results. Therefore, neurofeedback has the potential to provide low-cost yet holistic approach to the management of chronic pain.

**Keywords:** neurofeedback, EEG biofeedback, fMRI biofeedback, chronic pain, pain, fatigue, depression, anxiety, sleep

## 1. Introduction

Neurofeedback [1, 2] is a smart biofeedback platform which provides real-time feedback to individuals about their neurophysiological signals in order to achieve brain activity associated with therapeutic benefit. Brain activity of an individual is measured continuously using an EEG system during the course of neurofeedback training and parameters describing neurophysiological signals such as alpha power or peak alpha frequency are calculated in real-time [3]. These calculated features of ongoing brain activity are then presented to the individual either in an audio or visual form [3]. The idea behind this is that through repeated provision of such feedback, the individual gains an awareness of their current brain state and can identify different mental strategies which help them achieve the desired brain state [4]. Once the individual identifies strategies which work for them, they can keep practicing them over the course of multiple sessions with the final aim of being able to implement these strategies independent of a neurofeedback session.

Neurofeedback has already been investigated extensively for the management of several neuropsychiatric conditions [5] such as Attention Deficit Hyperactivity Disorder (ADHD) [6], depression and anxiety [7], cognition [8] and stroke

 IntechOpen

rehabilitation [9] for example. Being able to target brain signals through neurofeedback can be of great benefit in conditions such as chronic pain. This is because the perception of chronic pain depends on how multiple regions of the brain process the ascending pain signals [10, 11]. Such central processing of incoming pain signals has been shown to be different in chronic pain patients compared to healthy participants by a number of studies [12–14]. Considering the brain plays such an important in the development and maintenance of chronic pain state, being able to target changes in the neurophysiological signal which reflect such brain activity using a novel therapy such as neurofeedback is of great interest.

The field of neurofeedback therapy for chronic pain is rapidly developing. Several studies have been performed on a range of medical conditions over the last decade [15]. The current studies are highly heterogenous with a number of variations in neurofeedback protocol and delivery [15, 16]. This chapter aims to give an overview of the neurophysiological changes observed in chronic pain and how these have been targeted by different neurofeedback studies. We also discuss the different aspects of neurofeedback protocols which have been used so far and the outcomes of these studies in terms of reduction in pain and pain associated symptoms.

## 2. Neural pathways underlying pain perception

Our understanding of the neuroscience underlying pain has evolved significantly over time. Neural pathways involved in pain perception have been shown in **Figure 1**. One of the earliest theories explaining pain was the "specificity theory" (**Figure 1**: Red pathway). According to this theory, pain is experienced when an injury to a particular part of the body leads to signals being relayed via nociceptive neurons to the "pain center" [17]. The brain was considered to be a "passive recipient of sensory information" [17].

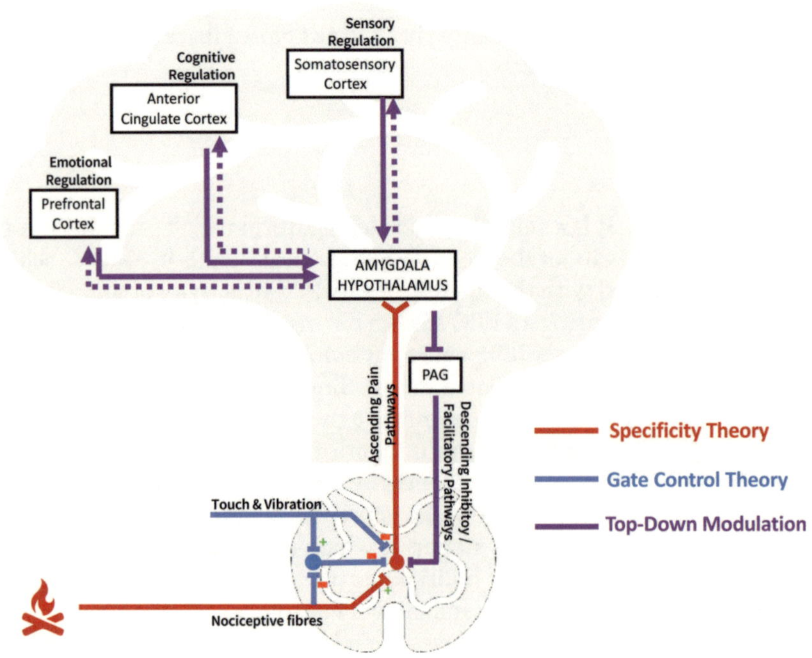


**Figure 1.**
*Neural pathways underlying pain perception proposed by different pain theories.*

One of the landmark theories which was highly influential in changing this prior understanding of pain was the Gate Control Theory by Melzack and Wall (1965) [18] (**Figure 1**: Blue pathway). This theory proposed that several neurons in the spinal cord, such as large fibers carrying touch and vibration sensations as well as interneurons in substantia gelatinosa of the dorsal horn, modulate the incoming signals from the site of pain, thereby influencing the final signal which is transmitted to the brain for processing.

Since then, advances in neuroimaging has revealed that in addition to neural pathways in the spinal cord, several cortical structures are also involved in modulating pain perception [19, 20] (**Figure 1**: Purple pathway). Some of the areas which have been reported to be involved include anterior cingulate gyrus, somatosensory cortex, insular cortex, thalamus and prefrontal cortex [19]. These findings suggest that there is not a single "pain centre". Instead, pain is processed by a "pain matrix" connecting different parts of the brain, thereby, reinforcing the idea that pain perception is a result of several sensory, affective and cognitive processes [10, 11]. Therefore, pain experienced by an individual is an integration of the current information about the painful sensory stimulation and prior information from previous experiences which influence the emotions, anxiety, attention and expectations of the individual about the pain [21].

Different areas of the cortex constituting the pain matrix project onto the hypothalamus and amygdala, which then give rise to both descending inhibitory pathways and descending facilitatory pathways [21, 22]. These descending pathways directly project onto the dorsal horn of the spinal cord where gating of pain is occurring, therefore, influence the signals which are relayed up the ascending pathways [21, 22]. This process is known as top-down modulation of pain [23].

In summary, the pain perceived by an individual is an integration of how different parts of the cortex process the ascending pain signals as well as how the activity of these cortical and subcortical structures influence the ascending pain signals via descending pathways [17, 21, 24]. With the discovery of these higher-order processes which influence pain perception, several neuromodulatory therapies such as neurofeedback (NFB), hypnosis and meditation, have been explored with the potential of controlling pain by influencing this supraspinal cortical processing of pain [25].

## 3. Brain rhythms associated with chronic pain

Generally, the EEG oscillations are categorized based on their frequency into theta (4–7 Hz), alpha (8–12 Hz), low beta or $beta_1$ (15–20 Hz) and high beta or $beta_2$ (22–30 Hz) [26–29]. Another oscillation which is widely investigated in the field of neuromodulation is sensorimotor rhythm (SMR). SMR refers to oscillations in the 12–15 Hz range which appear in spindle-like pattern over the sensorimotor cortex during idling of the motor cortex [30, 31]. Motor execution or motor imagery which activates the motor cortex leads to a decrease in the SMR activity [31].

Each of these brain rhythms is associated with a specific cognitive state. For instance, whilst alpha waves have been associated with a relaxed state, beta waves are associated with wakefulness and a state of engagement in task. Theta waves have been associated with drowsiness [27, 32].

Patients with chronic pain have differences in their resting-state brain (EEG) oscillations from healthy individuals. An example of a chronic pain condition which has been extensively investigated for identification of EEG correlates of chronic pain has been spinal cord injury (SCI). A study by Sarnthein et al. [12] showed that SCI patients with central neuropathic pain had increased activity of theta and beta oscillations compared to healthy individuals. These findings were confirmed by another study [13] which observed similar increases in theta and beta activity, but in addition, also identified lower levels of alpha activity in this patient population. This association between chronic pain and EEG changes was further strengthened when Jensen et al. [33] demonstrated that even within a group of patients with spinal cord injury, individuals with central neuropathic pain had higher theta and lower alpha activity than patients with spinal cord injury but no chronic pain.

These patterns of EEG have also been reported in other chronic pain conditions. For instance, patients with migraine have higher theta and delta power compared to healthy controls [14]. Patients with fibromyalgia have been shown to have higher theta activity with sources estimated to be in the left dorsolateral prefrontal and orbitofrontal cortex, higher beta and gamma activity with sources estimated to be in the insular, primary motor and primary and secondary somatosensory cortices and slowing of the dominant alpha peak [34].

Identification of such neurophysiological correlates of chronic pain is important as it not only provides the necessary feedback signal to increase voluntary control in therapies such as neurofeedback, but also allows monitoring the efficacy of the therapy in modulating the neurophysiological processes targeted by the therapy.

## 4. Neurofeedback training protocols

There are two key modalities which have been used to provide neurofeedback – EEG neurofeedback and fMRI neurofeedback. Whilst EEG neurofeedback provides feedback based on the neurophysiological signals recorded through an EEG system, fMRI neurofeedback provides feedback based on the degree of activation of a particular region of the brain detected using fMRI imaging in real time [35]. Hence it is inevitable that there is some lag between the activation and signal detection in fMRI neurofeedback which happens almost instantaneously in EEG neurofeedback [35].

Evidence regarding efficacy of fMRI neurofeedback in pain is limited understandably due to the increased difficulties and expenses associated with this form. The common region of interest which has been targeted in fMRI studies has been rostral anterior cingulate gyrus (rACC), whereby increased activity of rACC, measured through detecting an increase in blood oxygen level dependent signal from the region, has been associated with pain reduction [36, 37]. However, these studies have been severely limited in terms of number of sessions [37, 38], therefore the full benefit of the neurofeedback which occurs over the course of several sessions has not been explored yet in fMRI neurofeedback for chronic pain.

A number of brain rhythms have been targeted by EEG neurofeedback in order to increase resilience to pain (**Table 1**). The commonly targeted rhythms include theta (4–7 Hz), alpha (8–13 Hz), beta (14–30 Hz) and sensorimotor (12–15 Hz over the sensorimotor area) [17]. However, the change desired in each of these rhythms varies. Whilst pain reduction has been associated with an increase in the power of alpha and sensorimotor rhythms, contrastingly, a decrease in theta and beta rhythms have been associated with pain relief [17]. However, very few studies target these signals in isolation [20, 39]. More often studies target multiple signals at the same time, whereby patients are either shown each rhythm individually at the same time or they are shown feedback based on the ratio of two such signals [40–43].

In general, neurofeedback sessions tend to be 30–45 minutes long and patients are offered 20–40 sessions [15]. The frequency of these sessions ranges from one to five times a week, but studies which administered more frequent sessions have reported greater pain relief. Commonly used electrodes for providing feedback include C3, C4, Cz, T3, T4, FP1, P3 and P4 [15].

| Brain rhythm | Frequency | Desired change |
|---|---|---|
| Theta | 4–7 Hz | Decrease in power |
| Alpha | 8–13 Hz | Increase in power |
| Beta | 14–30 Hz | Decrease in power |
| Sensorimotor rhythm | 12–15 Hz<br>Over sensorimotor cortex | Increase in power |

**Table 1.**
*Neurofeedback targets [17].*

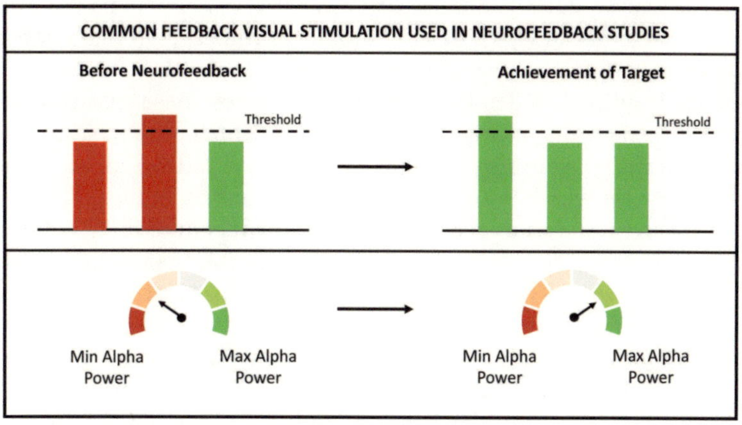


**Figure 2.**
*Schematic representation of visual stimulus provided in different neurofeedback studies.*

Feedback has been provided in a range of ways. Auditory feedback has been mainly in the form of changing volume of sound, whereby, achievement of signal has been associated with an increase in the volume heard [44]. Visual feedback used has been more varied (**Figure 2**). Some studies use simple bars to show the feedback, whereby the height of the bar is proportional to the intensity of the signal [45]. Other studies have changed the color of the bar on achievement of signal such that when the threshold is met, the color turns green, otherwise it remains red [43]. Some studies have tried to engage the users through the idea of games whereby the width of a river increases as the intensity of signal increases for instance [41, 46]. Therefore, feedback has been provided in a range of ways. Another form of stimulation which can be explored in the context of neurofeedback is tactile stimulation. Some studies have even combined two forms of stimuli such as visual and auditory whereby patient is shown an angry and shouting patient [36]. In order to calm the patient, the individual has to achieve the desired changes in the brain rhythms.

## 5. Efficacy of neurofeedback in management of chronic pain

Several neurofeedback studies have shown pain reduction following neurofeedback. Key randomized controlled trials in the field have been summarized in **Table 2**. Reduction in pain has been reported across several pain conditions such as Fibromyalgia [27, 29, 36, 41], Central Neuropathic Pain in Paraplegic patients [28, 43, 47–49], Traumatic Brain Injury [39, 50], Chemotherapy-Induced Peripheral Neuropathy [51], Primary Headache [52], Complex Regional Pain Syndrome Type I [53], Post-Herpetic Neuralgia [37] and chronic lower back pain [54]. There is a wide range of pain reduction reported which can range from an average of 6–82% reduction in pain intensity [15]. A recent systematic review published showed that ten out of twenty-one studies published in the field reported a pain reduction of greater than 30% which is considered to be clinically significant reduction [15].

Such variability in the degree of pain reduction could be due to a number of aspects of the neurofeedback protocol ranging from number of sessions, frequency of sessions, target frequencies and electrodes used for feedback, for example. The neurofeedback studies conducted so far have been highly variable on more than one of these aspects [15, 16], making comparison of results across studies impossible. Therefore, it is difficult to determine which of these parameters is responsible for the difference or how to best optimize each of these aspects of the training.

Most of the neurofeedback studies have measured changes in pain immediately following neurofeedback [39, 43, 52, 55, 56]. Furthermore, pain reduction has been reported to be sustained even at follow up of 3–6 months after completion of neurofeedback training [28, 36, 41, 49–51, 54]. However, these studies do not report whether the corresponding change in brain rhythm which were measured following completion of training were also sustained at long-term follow-up. We do not know the length of time for which the effect of neurofeedback on brain rhythms is sustained. Interestingly, one study reported that although pain reduction did not occur immediately following completion of the training course, there was improvement in pain at follow-up [36]. This could suggest that perhaps NFB could lead to changes in the underlying brain networks which occurs over a longer period of time but can be sustained for longer duration. These results provide the preliminary evidence for potential of neurofeedback for providing analgesia in chronic pain.

It has been shown that neurofeedback not only leads to reduction in pain but leads to improvement in a number of pain associated symptoms such as depression

| Study | Chronic pain condition | Target brain oscillation | % Pain reduction | Pain associated symptoms reported to improve following NFB |
|---|---|---|---|---|
| Goldway et al. (2019) [36] | Fibromyalgia | ↓Amygdala activation (fMRI) | 7% | REM latency<br>Sleep quality |
| Prinsloo et al. (2018) [50] | Chemotherapy-induced peripheral neuropathy | ↑Alpha<br>↓Beta | 45% | Fatigue<br>Cancer-related symptoms<br>Physical functioning<br>Quality of life |
| Guan et al. (2015) [37] | Post-herpetic Neuralgia | ↓rACC activity (fMRI) | 64% | None studied |
| Farahani et al. (2014) [45] | Primary headache | ↑SMR<br>↓Theta<br>↓Beta | 19% | None studied |
| Caro et al. (2011) [29] | Fibromyalgia | ↑SMR<br>↓Theta<br>↓Beta | 39% | Fatigue |
| Kayiran et al. (2010) [40] | Fibromyalgia | ↑SMR<br>↓Theta<br>↓Beta | 82% | Fatigue<br>Depression<br>Anxiety<br>Social functioning<br>Physical functioning |

**Table 2.**
*Randomized controlled trials investigating role of neurofeedback in chronic pain conditions.*

[27, 39, 41, 54, 57–60], anxiety [27, 41, 54, 57, 59], fatigue [27, 29, 41, 49, 51], and sleep [36, 39, 49–51, 57]. These symptoms have been known to co-exist with pain in chronic pain conditions and also known to exacerbate the individual's pain on a day-to-day basis [61–63]. Therefore, by being able to target these symptoms along with pain, neurofeedback has the potential to holistically improve the well-being of these individuals. A summary of different symptoms which have been shown to improve following neurofeedback have been shown in **Figure 3**.

Current neurofeedback studies have a number of limitations. There are currently only seven controlled trials in the field [29, 41, 47, 51, 52, 64, 65], of which only one trial is of high quality [65]. Most of the trials lack appropriate blinding as the control group are often patients on other pain medications [29, 66]. This makes the blinding of patient difficult and could lead to patient's belief in treatment affecting the results. Only two studies have implemented sham neurofeedback [36, 37].

The best sham treatment to offer is debatable. One would argue that patients could be shown the feedback signal from another region of the brain. But this might not be best as it might be the case that another region which is used for feedback might be the undiscovered part of the pain matrix. Another way to provide sham feedback would be to show the individual the recording from another participant or their own recording in a reverse order. Whilst this might be a true sham condition as the feedback shown to the individual would be independent of the individual's brain activity, it might mean that the patients find no relief of symptoms and discover that it is a sham treatment. Either way, such sham neurofeedback needs to be implemented by more studies in order to truly understand whether the pain reduction reported in these individuals is due to underlying changes in neuronal networks.

Whilst we have learnt a lot about neurofeedback over the past decade, there is still a lot which is unknown about this technique. Neurofeedback differs from other neuromodulatory techniques such entrainment and transcranial magnetic stimulation in that neurofeedback involves active involvement of the individuals in changing

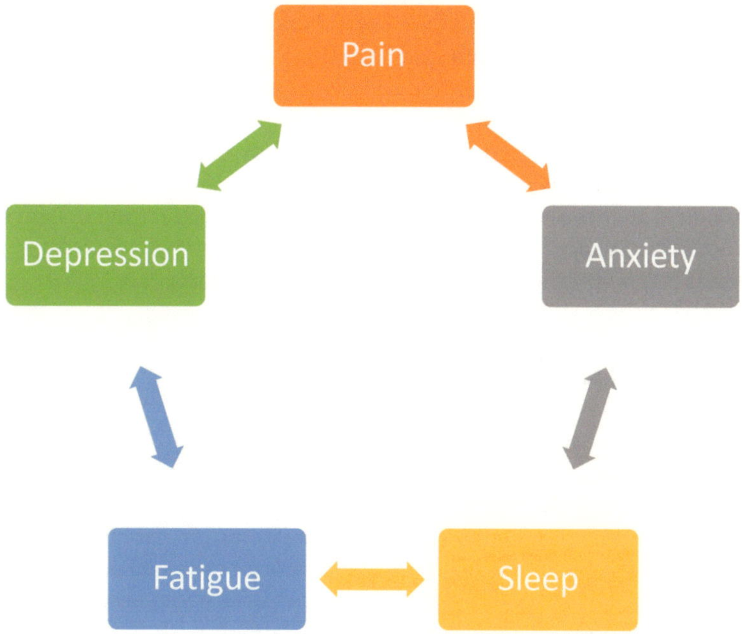


**Figure 3.**
*Schematic representation of pain and pain associated symptoms in chronic pain syndromes.*

the brain oscillations, as opposed to passive reception of stimulation [5]. We do not know which of these is a more efficient technique to alter brain oscillations yet. Furthermore, it is also unknown what mental strategies in particular are associated with changes in brain oscillations seen in the studies so far. Some of the common instructions given to patients undergoing training involve asking them to stay relaxed, imagining happy moments, revisiting happy memories and thinking about favorite family member or friends. However, none of the studies so far document which of these strategies actual work for the patients. Therefore, further qualitative studies are required to see what patients have been using to actively change their brain oscillations during neurofeedback in order to provide more focused instructions to patients undergoing training. Furthermore, studies should aim to analyze the correlation between neurophysiological signal and pain reduction rather than solely focusing on the behavioral outcomes [29, 41, 47, 51, 52, 64, 65]. Establishing such correlation between behavioral change and changes in neurophysiological signal is key to understanding whether the pain relief is truly due to neurofeedback.

In addition to this, there is also a possibility that once the patients have been able to identify the mental strategy which allows them to achieve the desired brain state and practice in the neurofeedback setting for a number a sessions, they might be able to implement such mental strategies without the ongoing EEG signal feedback. It is not clear if this possible or how long it might take for an individual to become independent of the EEG feedback and still receive pain relief.

The current neurofeedback studies are highly heterogenous. It is unclear which brain regions, oscillations, feedback form or training length is required to optimize the improvement in pain. More studies are required comparing one aspect of the neurofeedback training program at once in order to determine which of these parameters provide the most therapeutic benefit.

Another area of uncertainty is the efficacy of neurofeedback in different pain conditions. Studies so far have shown that all chronic pain condition report pain reduction to some degree following neurofeedback. However, it is not known whether neurofeedback is better for some chronic pain conditions than others. It might be the case that neuronal changes seen following neurofeedback is linked to central sensitization only, in which case several chronic pain conditions may benefit from it equally as many pain conditions have this as the underlying pathology. However, we do not know whether it is equally as good at treating nociceptive pain as seen in conditions such as arthritis.

Furthermore, the role that neurofeedback will play in pain management in the future is not clear [16]. It is not clear whether it has the true potential to substitute pharmacological agents completely. It might be the case that it might reduce the escalation of opioid usage in this patient cohort. Hence further studies are needed to determine the maximum potential of this form of therapy.

## 6. Adverse effects associated with Neurofeedback

In general, neurofeedback is well tolerated with a minority of patients experiencing mild adverse events. These adverse events are often self-limiting and tend to be controlled by decreasing the frequency of training [43, 48]. Adverse events seen in neurofeedback studies seem to be more common in certain patient groups than others. For instance, some individuals with spinal cord injury and central neuropathic pain have reported some hypersensitivity of soles of the feet due to recovery of proprioception or spasms of the lower limb, [28, 48]. Patients with traumatic brain injury have reported an increase in nausea and the intensity of their headaches [39, 67]. It is difficult to confirm that these side-effects are due to NFB as these reported symptoms are often seen in these conditions irrespective of provision of neurofeedback therapy. Overall, NFB is safe and well-tolerated in majority of patients in most clinical studies.

## 7. Delivery of home-based neurofeedback therapy

Neurofeedback has also been delivered in the home setting by a few recent studies [43, 48]. This can be achieved through the use of a headset which records activity from one single electrode, such as C4 [43, 48] or FP1 [39] and makes use of an app on tablets to analyze and showcase feedback to the individual [28, 48]. Such systems have been implemented in patients with central neuropathic pain [43, 48] as well as traumatic brain injury [39]. Patients could practice neurofeedback for 5- or 10-minutes sessions as and when they wanted.

These studies have shown some promising results. With further expansion of this technology, it might be possible for individuals to benefit from neurofeedback at their home as and when required as patients have on average used neurofeedback 3–40 times over the course of 2–3 months in these studies [43, 48]. Two of these studies have reported around 33% reduction in pain [43, 48] whereas one of them reported 16% reduction in pain [39] on average in participants who tried these home-based systems.

One of these studies also performed qualitative research on user experience following such home-based systems [43]. Overall, it was reported that the patient satisfaction score was high when measured using QUESB (Quebec User Evaluation of Satisfaction Questionnaire). According to the patients, the key factors which affected the frequency of their use of the home-based device were their health state, availability of free time and their intensity of pain. Patients also put effectiveness, ease of use and comfort as their main priority when using any such home-based device. Hence whilst the current home-based technology used in this study showed that it could record the data with decent quality, it also highlighted that patients wanted technology which was able to provide neurofeedback wirelessly using headset and smart device as well as collect information from the scalp without the use of gel to connect electrodes.

Being able to do this on a regular basis would also increase the efficacy of the therapy and patients might be able to use neurofeedback in addition to or instead of commonly used pharmacological agents which are associated with significant adverse effect profiles. Therefore, home-based neurofeedback can act as a novel treatment option to provide pain relief to patients with much fewer side effects than current pharmacological agents [68].

## 8. Conclusions

Neurofeedback is a newly emerging technique which can be used to achieve brain states associated with increased resilience to pain. The results so far have been very promising not only in terms of improvement in chronic pain, where as many as half of the studies in the field have shown clinically significant reduction in clinical pain following neurofeedback, but also in terms of improvement in pain associated symptoms such as fatigue, depression, anxiety and sleep which have also been reported to improve with neurofeedback. Being able to target all of these co-morbidities holistically using neurofeedback is key for the overall improvement in the well-being of chronic pain patients because these factors are often interlinked and aggravate each other.

There is still a lot of work that needs to be done. Different aspects of training protocols, such as target signal, number of sessions, length of sessions and scalp region of interest, need to be optimized in order to identify parameters which lead not only to successful modulation of the brain activity but also a corresponding change in pain signals. Currently, it is not clear what neurofeedback protocol brings about maximum pain relief for patients.

Furthermore, identification of mental strategies which enable individuals to reach therapeutic brain states is also required, with the aim being that eventually individuals will be able to practice these strategies independent of the feedback system after an initial course of training sessions. Whilst, there is a lot of work to do, the results so far have been promising, opening window of opportunity to manage a number of chronic pain conditions at low cost and without the side effects associated with the currently available pharmacological agents.

## Conflict of interest

The authors have no conflict of interest to declare.

## Author details

Kajal Patel[1*], Manoj Sivan[2,3], James Henshaw[2] and Anthony Jones[2]

1 School of Medicine, University of Manchester, Manchester, UK

2 The Human Pain Research Group, Division of Neuroscience and Experimental Psychology, University of Manchester, Manchester, UK

3 Academic Department of Rehabilitation Medicine, University of Leeds, Leeds, UK

*Address all correspondence to: kj.patel1020@gmail.com

## References

[1] Evans JR. Handbook of neurofeedback: dynamics and clinical applications. CRC Press. 2007;

[2] Marzbani H, Marateb HR, Mansourian M. Methodological note: Neurofeedback: A comprehensive review on system design, methodology and clinical applications. Basic Clin Neurosci. 2016 Mar 1;7(2):143-58.

[3] Bagdasaryan J, Le Van Quyen M. Experiencing your brain: Neurofeedback as a new bridge between neuroscience and phenomenology. Front Hum Neurosci. 2013;7(OCT):1-10.

[4] Alkoby O, Abu-Rmileh A, Shriki O, Todder D. Can We Predict Who Will Respond to Neurofeedback? A Review of the Inefficacy Problem and Existing Predictors for Successful EEG Neurofeedback Learning. Vol. 378, Neuroscience. Elsevier Ltd; 2018. p. 155-64.

[5] Budzynski TH, Budzynski HK, Evans JR, Abarbanel A. Introduction to quantitative EEG and neurofeedback: Advanced theory and applications. Acad Press. 2009;

[6] Van Doren J, Arns M, Heinrich H, Vollebregt MA, Strehl U, K. Loo S. Sustained effects of neurofeedback in ADHD: a systematic review and meta-analysis. European Child and Adolescent Psychiatry. 2019.

[7] Schoenberg PLA, David AS. Biofeedback for psychiatric disorders: A systematic review. Appl Psychophysiol Biofeedback. 2014;39(2):109-35.

[8] Gruzelier JH. EEG-neurofeedback for optimising performance. I: a review of cognitive and affective outcome in healthy participants. Neuroscience and Biobehavioral Reviews. 2014. p. 124-41.

[9] Carvalho R, Dias N, Cerqueira JJ. Brain-machine interface of upper limb recovery in stroke patients rehabilitation: A systematic review. Physiotherapy Research International. 2019.

[10] Katz WA, Rothenberg R. The nature of pain: Pathophysiology. J Clin Rheumatol. 2005;11(2):11-5.

[11] Reddan MC, Wager TD, M.C. R. Brain systems at the intersection of chronic pain and self-regulation. Neurosci Lett [Internet]. 2019;702:24-33. Available from: http://www.elsevier.com/locate/neulet

[12] Sarnthein J, Stern J, Aufenberg C, Rousson V, Jeanmonod D. Increased EEG power and slowed dominant frequency in patients with neurogenic pain. Brain. 2006;129(1):55-64.

[13] Boord P, Siddall PJ, Tran Y, Herbert D, Middleton J, Craig A. Electroencephalographic slowing and reduced reactivity in neuropathic pain following spinal cord injury. Spinal Cord. 2008;

[14] Dos Santos Pinheiro ES, De Queirós FC, Montoya P, Santos CL, Do Nascimento MA, Ito CH, et al. Electroencephalographic patterns in chronic pain: A systematic review of the literature. PLoS One. 2016;11(2):1-26.

[15] Patel K, Sutherland H, Henshaw J, Taylor JR, Brown CA, Casson AJ, et al. Effects of neurofeedback in the management of chronic pain: A systematic review and meta-analysis of clinical trials. Eur J Pain (United Kingdom) [Internet]. 2020 Jun 5 [cited 2020 Jun 8];24:1440-1457. Available from: https://onlinelibrary.wiley.com/doi/abs/10.1002/ejp.1612

[16] Roy R, de la Vega R, Jensen MP, Miró J. Neurofeedback for Pain Management: A Systematic Review. Frontiers in Neuroscience. 2020.

[17] Jensen MP, Sherlin LH, Hakimian S, Fregni F. Neuromodulatory approaches for chronic pain management: Research findings and clinical implications. J Neurother. 2009;13(4):196-213.

[18] Melzack R, Wall PD. Pain Mechanisms : A New Theory. Science (80- ) [Internet]. 1965 [cited 2019 Sep 29];150:971-8. Available from: https://science-sciencemag-org.manchester.idm.oclc.org/content/sci/150/3699/971.full.pdf

[19] Apkarian AV, Bushnell MC, Treede R-D, Zubieta J-K. Human brain mechanisms of pain perception and regulation in health and disease. Eur J Pain [Internet]. 2005 [cited 2019 Sep 29];9:463-84. Available from: www.EuropeanJournalPain.com

[20] Mayaud L, Wu H, Barthélemy Q, Favennec P, Delpierre Y, Congedo M, et al. Alpha-phase synchrony EEG training for multi-resistant chronic low back pain patients: an open-label pilot study. Eur Spine J. 2019;

[21] Ossipov MH, Dussor GO, Porreca F. Central modulation of pain. J Clin Invest [Internet]. 2010 [cited 2019 Sep 30];120(11):3779-87. Available from: http://www.jci.org

[22] DeLeo JA. Basic science of pain. J Bone Jt Surg. 2006;88(2):58-62.

[23] Donaldson LF, Lumb BM. Top-down control of pain [Internet]. Vol. 595, Journal of Physiology. 2017 [cited 2020 Jan 12]. p. 4139-40. Available from: https://www.ncbi.nlm.nih.gov/pmc/articles/PMC5491874/

[24] Apkarian A V, Hashmi JA, Baliki MN. Pain and the brain: Specificity and plasticity of the brain in clinical chronic pain. [cited 2019 Nov 9]; Available from: http://www.iasp-pain.org/

[25] Jensen MP, Day MA, Miro J. Neuromodulatory treatments for chronic pain: efficacy and mechanisms. Nat Rev Neurol [Internet]. 2014;10(3):167-78. Available from: http://ovidsp.ovid.com/ovidweb.cgi?T=JS&PAGE=reference&D=med10&NEWS=N&AN=24535464

[26] Kropotov JD. Rhythms of the Healthy Brain. In: Functional Neuromarkers for Psychiatry [Internet]. 2016 [cited 2020 Jan 10]. p. 403-5. Available from: http://dx.doi.org/10.1016/B978-0-12-410513-3.00038-3

[27] Kayiran S, Dursun E, Ermutlu N, Dursun N, Karamursel S. Neurofeedback in fibromyalgia syndrome. Agri. 2007;19(3):47-52.

[28] Hassan MA, Fraser M, Conway BA, Allan DB, Vuckovic A. The mechanism of neurofeedback training for treatment of central neuropathic pain in paraplegia: A pilot study. BMC Neurol. 2015;15(1).

[29] Caro XJ, Winter EF. EEG biofeedback treatment improves certain attention and somatic symptoms in fibromyalgia: A pilot study. Appl Psychophysiol Biofeedback [Internet]. 2011 Sep [cited 2019 Oct 22];36(3):193-200. Available from: http://ovidsp.ovid.com/ovidweb.cgi?T=JS&PAGE=reference&D=emed12&NEWS=N&AN=51466220

[30] Collura TF, Siever D. Audio-Visual Entrainment in Relation to Mental Health and EEG. In: Introduction to Quantitative EEG and Neurofeedback. 2009.

[31] Timmers D. Treating Attention Deficits and Impulse Control. In: Clinical Neurotherapy: Application of Techniques for Treatment. 2013.

[32] Jensen MP, Hakimian S, Sherlin LH, Fregni F, M.P. J, S. H, et al. New Insights Into Neuromodulatory Approaches for the Treatment of Pain. J Pain [Internet].

2008 [cited 2019 Sep 29];9(3):193-9. Available from: http://ovidsp.ovid.com/ovidweb.cgi?T=JS&PAGE=reference&D=emed10&NEWS=N&AN=50020976

[33] Jensen MP, Sherlin LH, Gertz KJ, Braden AL, Kupper AE, Gianas A, et al. Brain EEG activity correlates of chronic pain in persons with spinal cord injury: Clinical implications. Spinal Cord. 2013;51(1):55-8.

[34] Lim M, Kim JS, Kim DJ, Chung CK. Increased low-and high-frequency oscillatory activity in the prefrontal cortex of fibromyalgia patients. Front Hum Neurosci. 2016;10(MAR2016):1-11.

[35] Hawkinson JE, Ross AJ, Parthasarathy S, Scott DJ, Laramee EA, Posecion LJ, et al. Quantification of adverse events associated with functional MRI scanning and with real-time fMRI-based training. Int J Behav Med [Internet]. 2012;19(3):372-81. Available from: http://ovidsp.ovid.com/ovidweb.cgi?T=JS&PAGE=reference&D=med8&NEWS=N&AN=21633905

[36] Goldway N, Ablin J, Lubin O, Zamir Y, Keynan JN, Or-Borichev A, et al. Volitional limbic neuromodulation exerts a beneficial clinical effect on Fibromyalgia. Neuroimage [Internet]. 2019;186:758-70. Available from: http://ovidsp.ovid.com/ovidweb.cgi?T=JS&PAGE=reference&D=medl&NEWS=N&AN=30408596

[37] Guan M, Ma L, Li L, Yan B, Zhao L, Tong L, et al. Self-regulation of brain activity in patients with postherpetic neuralgia: a double-blind randomized study using real-time FMRI neurofeedback. PLoS One [Internet]. 2015 Apr;10(4):e0123675. Available from: http://www.plosone.org/article/fetchObject.action?uri=info:doi/10.1371/journal.pone.0123675&representation=PDF

[38] Goldway N, Ablin J, Lubin O, Zamir Y, Jacob Keynan N, Or-Borichev A, et al. Beyond pain in fibromyalgia: Limbic related eeg-neurofeedback training improves sleep and affect. Biol Psychiatry [Internet]. 2018 May;83(9 Supplement 1):S447. Available from: http://ovidsp.ovid.com/ovidweb.cgi?T=JS&PAGE=reference&D=emexa&NEWS=N&AN=621901919

[39] Elbogen EB, Alsobrooks A, Battles S, Molloy K, Dennis PA, Beckham JC, et al. Mobile Neurofeedback for Pain Management in Veterans with TBI and PTSD. Pain Med [Internet]. 2019;0(0):1-9. Available from: https://academic.oup.com/painmedicine/advance-article/doi/10.1093/pm/pnz269/5614403

[40] S. P, T. K, E. B, D. D, D. N. Neuromodulation as a treatment for chemotherapy-induced peripheral neuropathy (CIPN): Examining differences between placebo and neurofeedback. Psychosom Med [Internet]. 2019;81(4):A63. Available from: http://ovidsp.ovid.com/ovidweb.cgi?T=JS&PAGE=reference&D=emexb&NEWS=N&AN=627783404

[41] Kayıran S, Dursun E, Dursun N, Ermutlu N, Karamürsel S. Neurofeedback intervention in fibromyalgia syndrome; a randomized, controlled, rater blind clinical trial. Ahles Alanoglu, Anderberg, Babu, Basmajian, Bennett, Caruso, Carville, Chiarioni, Corapcoglu, Crider, Da Costa, Denk, Draizar, Dursun, Dursun, Dursun, Egner, Egner, Egner, Gentil, Grace, Gracely, Guedj, Hidalgo, Hisli, Ho, Howe, Intiso, Jensen, Jensen, r A, editor. Appl Psychophysiol Biofeedback [Internet]. 2010;35(4):293-302. Available from: http://ovidsp.ovid.com/ovidweb.cgi?T=JS&PAGE=reference&D=med7&NEWS=N&AN=20614235

[42] Vuckovic A, Hasan MA, Fraser M, Conway B, Allan DB. A Pilot Study on Clinical and Neurological Effects of Neurofeedback Training for Treatment of Central Neuropathic

Pain. In: Jensen, W and Andersen, OK and Akay M, editor. REPLACE, REPAIR, RESTORE, RELIEVE - BRIDGING CLINICAL AND ENGINEERING SOLUTIONS IN NEUROREHABILITATION [Internet]. GEWERBESTRASSE 11, CHAM, CH-6330, SWITZERLAND: SPRINGER INT PUBLISHING AG; 2014. p. 823-31. (Biosystems and Biorobotics; vol. 7). Available from: http://link.springer.com/10.1007/978-3-319-08072-7_113

[43] Al-Taleb MKH, Purcell M, Fraser M, Petric-Gray N, Vuckovic A. Home used, patient self-managed, brain-computer interface for the management of central neuropathic pain post spinal cord injury: Usability study. J Neuroeng Rehabil [Internet]. 2019 [cited 2019 Nov 18];16(1). Available from: https://doi.org/10.1186/s12984-019-0588-7

[44] Abul Hassan M, Fraser M, Conway BA, Allan DB, Vuckovic A. The mechanism of neurofeedback training for treatment of central neuropathic pain in paraplegia: a pilot study. BMC Neurol. 2015 Oct;15.

[45] Vuckovic A, Hasan MA, Fraser M, Conway B, Allan DB. A Pilot Study on Clinical and Neurological Effects of Neurofeedback Training for Treatment of Central Neuropathic Pain. In: Jensen, W and Andersen, OK and Akay, M, editor. REPLACE, REPAIR, RESTORE, RELIEVE - BRIDGING CLINICAL AND ENGINEERING SOLUTIONS IN NEUROREHABILITATION. 2014. p. 823-31. (Biosystems and Biorobotics; vol. 7).

[46] Farahani DM, Tavallaie SA, Ahmadi K, Ashtiani AF. Comparison of neurofeedback and transcutaneous electrical nerve stimulation efficacy on treatment of primary headaches: A randomized controlled clinical trial. Iran Red Crescent Med J. 2014 Aug 1;16(8).

[47] Jensen MP, Sherlin LH, Askew RL, Fregni F, Witkop G, Gianas A, et al. Effects of non-pharmacological pain treatments on brain states. Clin Neurophysiol [Internet]. 2013 Oct [cited 2019 Oct 22];124(10):2016-24. Available from: http://ovidsp.ovid.com/ovidweb.cgi?T=JS&PAGE=reference&D=med9&NEWS=N&AN=23706958

[48] Vučković A, Altaleb MKH, Fraser M, McGeady C, Purcell M. EEG Correlates of Self-Managed Neurofeedback Treatment of Central Neuropathic Pain in Chronic Spinal Cord Injury. Front Neurosci [Internet]. 2019 Jul 25;13. Available from: https://www.frontiersin.org/article/10.3389/fnins.2019.00762/full

[49] Jensen MP, Gertz KJ, Kupper AE, Braden AL, Howe JD, Hakimian S, et al. Steps toward developing an EEG biofeedback treatment for chronic pain. Amtmann Bazanova, Boord, Bromm, Bromm, Bromm, Cardenas, Caro, Chen, Chen, Cook, Egner, Egner, Ehde, Finnerup, Gannon, Gevensleben, Gevensleben, Hammond, Hays, Huber, Jasper, Jensen, Jensen, Jensen, Jensen, Jensen, Kayiran, Krupp, Llinas, Raymond, Ros, Sa B, editor. Appl Psychophysiol Biofeedback [Internet]. 2013 Jun [cited 2019 Oct 22];38(2):101-8. Available from: http://ovidsp.ovid.com/ovidweb.cgi?T=JS&PAGE=reference&D=med9&NEWS=N&AN=23532434

[50] Hershaw JN, Hill-Pearson CA, Arango JI, Souvignier CAR, Pazdan CRM. Semi-Automated Neurofeedback Therapy for Persistent Postconcussive Symptoms in a Military Clinical Setting: A Feasibility Study. Mil Med [Internet]. 2020 Oct 11 [cited 2019 Oct 22]; Available from: http://www.ncbi.nlm.nih.gov/pubmed/31603218

[51] Prinsloo S, Novy D, Driver L, Lyle R, Ramondetta L, Eng C, et al. The Long-Term Impact of Neurofeedback on Symptom Burden and Interference in Patients With Chronic Chemotherapy-Induced Neuropathy: Analysis of a Randomized Controlled Trial. J Pain

Symptom Manage [Internet]. 2018 [cited 2019 Sep 29];55(5):1276-85. Available from: http://www.elsevier.com/locate/jpainsymman

[52] Farahani DM, Tavallaie SA, Ahmadi K, Fathi Ashtiani A, Sheikh M, Yahaghi E, et al. Comparison of Neurofeedback and Transcutaneous Electrical Nerve Stimulation Efficacy on Treatment of Primary Headaches: A Randomized Controlled Clinical Trial. Iran Red Crescent Med J [Internet]. 2014 Aug 5;16(7):e17799. Available from: http://ircmj.com/53808.pdf

[53] Jensen MP, Grierson C, Tracy-Smith V, Bacigalupi SC, Othmer S. Neurofeedback treatment for pain associated with complex regional pain syndrome type I. Turk BBBBCCCC deCharms EFHJMOOOPPQSS, editor. J Neurother [Internet]. 2007;11(1):45-53. Available from: http://ovidsp.ovid.com/ovidweb.cgi?T=JS&PAGE=reference&D=psyc5&NEWS=N&AN=2007-13570-004

[54] Mayaud L, Wu H, Barthélemy Q, Favennec P, Delpierre Y, Congedo M, et al. Alpha-phase synchrony EEG training for multi-resistant chronic low back pain patients: an open-label pilot study. Eur Spine J [Internet]. 2019 [cited 2019 Nov 18];28(11):2487-501. Available from: https://doi.org/10.1007/s00586-019-06051-9

[55] Caro XJ, Winter EF. EEG biofeedback treatment improves certain attention and somatic symptoms in fibromyalgia: a pilot study. Appl Psychophysiol Biofeedback [Internet]. 2011;36(3):193-200. Available from: http://ovidsp.ovid.com/ovidweb.cgi?T=JS&PAGE=reference&D=med7&NEWS=N&AN=21656150

[56] Vuckovic A, Altaleb MKH, Fraser M, McGeady C, Purcell M. EEG Correlates of Self-Managed Neurofeedback Treatment of Central Neuropathic Pain in Chronic Spinal Cord Injury. Front Neurosci. 2019 Jul;13.

[57] Ibric VL, Dragomirescu LG. Neurofeedback in pain management. In: Budzynski, TH and Budzynski, HK and Evans, JR and Abarbanel A, editor. Introduction to Quantitative EEG and Neurofeedback [Internet]. SARA BURGERHARTSTRAAT 25, PO BOX 211, 1000 AE AMSTERDAM, NETHERLANDS: Elsevier; 2009. p. 417-51. Available from: https://linkinghub.elsevier.com/retrieve/pii/B9780123745347000162

[58] Koberda JL. LORETA Z-Score Neurofeedback in Chronic Pain and Headaches. In: JF TR and L, editor. Z Score Neurofeedback [Internet]. 525 B STREET, SUITE 1900, SAN DIEGO, CA 92101-4495 USA: Elsevier; 2015. p. 115-39. Available from: https://linkinghub.elsevier.com/retrieve/pii/B9780128012918000066

[59] Koberda JL. Z-score LORETA neurofeedback as a potential therapy in depression/anxiety and cognitive dysfunction. Alonzo Choi, Kessler, Koberda, Koberda, Koberda, Koberda, Pittenger, Riva-Posse, Schlaepfer, Stevens, Thatcher, Williams A, editor. Z score neurofeedback Clin Appl [Internet]. 2015;93-113. Available from: http://ovidsp.ovid.com/ovidweb.cgi?T=JS&PAGE=reference&D=psyc12&NEWS=N&AN=2014-45849-005

[60] Koberda JL, Koberda P, Bienkiewicz AA, Moses A, Koberda L. Pain management using 19-electrode Z-score LORETA neurofeedback. Caro Ibric, Jensen, Jensen, Kayiran, Kayiran, Kenchadze, Koberda, Koberda, Koberda, Moisset, Moont, Prinsloo, Sawamoto, Sime, Stern, Stokes, Thatcher, Thatcher, Thatcher, Thatcher, Walker H, editor. J Neurother [Internet]. 2013;17(3):179-90. Available from: http://ovidsp.ovid.com/ovidweb.cgi?T=JS&PAGE=reference&D=psyc10&NEWS=N&AN=2013-30620-005

[61] Feingold D, Brill S, Goor-Aryeh I, Delayahu Y, Lev-Ran S. Depression and anxiety among chronic pain patients receiving prescription opioids and medical marijuana. J Affect Disord. 2017;

[62] Zis P, Daskalaki A, Bountouni I, Sykioti P, Varrassi G, Paladini A. Depression and chronic pain in the elderly: Links and management challenges. Clinical Interventions in Aging. 2017.

[63] Bonvanie IJ, Oldehinkel AJ, Rosmalen JGM, Janssens KAM. Sleep problems and pain: A longitudinal cohort study in emerging adults. Pain. 2016;

[64] N. G, J. A, O. L, Y. Z, J.N. K, A. O-B, et al. Volitional limbic neuromodulation exerts a beneficial clinical effect on Fibromyalgia. Neuroimage [Internet]. 2019;186:758-70. Available from: http://ovidsp.ovid.com/ovidweb.cgi?T=JS&PAGE=reference&D=medl&NEWS=N&AN=30408596

[65] M. G, L. M, L. L, B. Y, L. Z, L. T, et al. Self-regulation of brain activity in patients with postherpetic neuralgia: A double-blind randomized study using real-time fMRI neurofeedback. PLoS One [Internet]. 2015;10(4):e0123675. Available from: http://www.plosone.org/article/fetchObject.action?uri=info:doi/10.1371/journal.pone.0123675&representation=PDF

[66] S. K, E. D, N. E, N. D. Neurofeedback in fibromyalgia syndrome. Agri [Internet]. 2007;19(3):47-52. Available from: http://ovidsp.ovid.com/ovidweb.cgi?T=JS&PAGE=reference&D=emed10&NEWS=N&AN=350246563

[67] J.N. H, C.A. H-P, J.I. A, C.A.R. S. Semi-Automated Neurofeedback Therapy for Persistent Postconcussive Symptoms in a Military Clinical Setting: A Feasibility Study. Mil Med [Internet]. 2019; Available from: http://ovidsp.ovid.com/ovidweb.cgi?T=JS&PAGE=reference&D=emexc&NEWS=N&AN=629566784

[68] Luctkar-Flude M, Groll D. A Systematic Review of the Safety and Effect of Neurofeedback on Fatigue and Cognition. Integr Cancer Ther. 2015;14(4):318-40.

Section 6

# Privacy Security and Integrity of Data

Chapter 6

# Blockchain-Based Medical Record Management with Biofeedback Information

*Hui Li Wang, Shao-I Chu, Jiun-Han Yan, Yu-Jung Huang, I-Yueh Fang, Shu Ya Pan, Wei-Cheng Lin, Chao-Tien Hsu, Chih-Lung Hung, Tzung-Ching Lin and Te-Tsun Shen*

## Abstract

Blockchain is a new emerging technology of distributed databases, which guarantees the integrity, security and incorruptibility of data by means of the cryptography. Such features are suitable for secure and reliable data storage. This chapter investigates the blockchain-based architecture with applications to medical health record or biofeedback information management. This framework employs the smart contract to establish a medical record management system to ensure the privacy of patients. Moreover, the blockchain technique accelerates the medical record or information exchange such that the cost of human resource is significant reduced. All patients can manage their individual medical records and information easily in the different hospitals and clinics. They also have the privilege to deal with and authorize personal medical records in the proposed management framework.

**Keywords:** blockchain, medical record, security, privacy

## 1. Introduction

The conventional medical record systems face the complicated administration procedure for data processing to ensure patients' privacy, leading to the enormous waste of human resources. Such an architecture is obviously inefficient for the medical record exchange. Blockchain technique [1] has recently been adopted to secure medical data sharing and management. The cryptographic property in the blockchain networks guarantees the patients' privacy. Data integrity and incorruptibility protect medical data from being tampered. The blockchain can be viewed as a distributed database, which stores data in each network nodes to avoid the halting problem. It thus provides higher stability, consistency and attack-resistance. The problem of distributed denial-of-service attacks (DDOS) in the conventional centralized framework can be solved by the blockchain technique. Deployment of blockchain in the medical record system not only provides the reliable service but also speeds up the medical record exchange. Owing to decentralization, the ownership of the medical record is returned to the patients, allowing them to manage the medical record directly and take care of their own health.

Biofeedback is a technique that uses electrical sensors to measure human body functions such as blood pressure and heart rate. Biofeedback aims to help learn your body condition and how it works. You may have biofeedback training in clinics, medical centers and hospitals. These measurements or data about biofeedback are thus important for the future therapy. Such data storage or management is therefore required and may be integrated with the medical record system.

In this chapter, we investigate the blockchain-based medical record management systems in the literatures. The new medical record framework is thus proposed and implemented on an open-source platform, Ethereum. Note that Ethereum platform allows anyone to develop applications in blockchain networks. The Ethereum virtual machine (EVM) executes the smart contract on the virtual machine. With the aid of the smart contract, developers are able to put these codes on the blockchain. The codes will be automatically executed after the blockchain establishment. The medical record management system is developed by exploiting the smart contract. The aim of the level of health care will be achieved in the proposed architecture via the smart contracts.

## 2. Literatures surveys

Authors in [2] presented the MedRec system, a decentralized medical record management system based on blockchain technology. There are three types of Ethereum smart contracts to associate patients' medical data to allow third-party users to access the data. Yang *et al.* [3] further presented an attribute-based authentication mechanism on the MedRec system to enable the secure sharing of medical data. A high-level blockchain-based framework was designed in [4], where an identity-based authentication and key agreement protocol is applied to achieve user membership authentication. They also developed the MedShare [5] system to provide data provenance and control in cloud repositories among hospitals. Liang *et al.* [6] used the hyperledger fabric membership service and channel formation scheme to guarantee data privacy in a blockchain network for medical data sharing. The mobile application was also implemented to collect data from wearable devices for storage and sharing with healthcare providers. Patientory [7] is a peer-to-peer medical record data storage network. The software framework in [7] was presented to address the authentication, authorization, access control, data encryption interoperability enhancement and token management.

Authors in [8] proposed the MedChain system, where the timed-based smart contracts can interact with the various demands of health providers, patients and third parties. An incentive mechanism in [8] was also presented to leverage the degree of health providers about their efforts on maintaining medical records. In [9], an attribute-based signature scheme with multiple authorities was designed. There are multiple authorities without a centralized one to generate and deliver public/private keys of the patient, avoiding the escrow problem. Liu *et al.* [10] presented a healthcare insurance anti-fraud system based on blockchain. A hybrid architecture to facilitate access control of medical data was developed in [11]. A blockchain is used to manage identity and access control and acts as a tamper-proof log of access events. Hasavari *et al.* [12] introduced a combination of secure file transfer methods and blockchain techniques as a solution to record patient's emergency medical data such that ambulance crews can access and use it to provide high quality pre-hospital care.

## 3. Proposed Ethereum-based framework for medical record management

Instead of using the traditional centralized databases, the Ethereum-based blockchain is applied to our designed system framework of medical record management to ensure the security of data. The medical records are stored within individual nodes in the blockchain networks by utilizing the smart contracts. The automatic smart contracts for the administration procedure are also designed with an aim to reducing the waste of human resource and speeding up the medical process.

The presented medical record management system is essentially rooted in the Ethereum-based blockchain architecture. The management framework is developed and established based on the relationship among the smart contracts. The proposed architecture is modified from the framework in [2]. The whole system is viewed as a private blockchain network, where all medical records are stored to guarantee data security, privacy and integrity. Innovatively combined with the data exchange mechanism in [13], the user identity is directly recognized by the system and the corresponding privilege is authorized to ensure data integrity in the blockchain networks.

### 3.1 Blockchain-based medical record management system

**Figure 1** illustrates the proposed medical record management system with smart contracts. There are three types of smart contracts, including registrar contract, patient-provider relationship contract and summary contract.

#### *3.1.1 Registrar contract (RC)*

Similar to [2], this contract maps member identification strings to the Ethereum address identity. All the registered members are divided into two groups, patients and medical personnel. Each identity has different access rights for the proposed system.

Authorized privilege of patients:

a) Review their own medical records. b) Authorize their own medical records. Authorized privilege of the medical personnel:

a) Create/modify the authorized medical records. b) Review the authorized medical records. Notice that different kinds of the medical personnel has different authorized or restricted rights.

#### *3.1.2 Patient-provider relationship contract (PPR)*

Each PPR smart contract is a medical certificate. The PPR smart contract is utilized to record the current situation of medical records, details of diagnosis and the access permission of different summary contracts. Sometimes other relevant PPR diagnostic address information will also be included. All summary smart contracts must be licensed by the owner of the PPR contract to access the PPR smart contract.

As shown in **Figure 2**, the medical personnel will be allowed to modify or read only based on the access right in the summary contracts after getting the permits. Moreover, the administration system can track the current diagnosis by the assistance of the PPR contract.

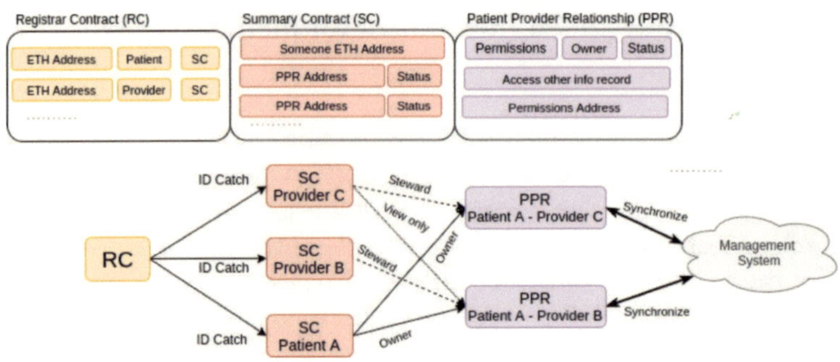


**Figure 1.**
*Proposed smart contract-based medical management system.*

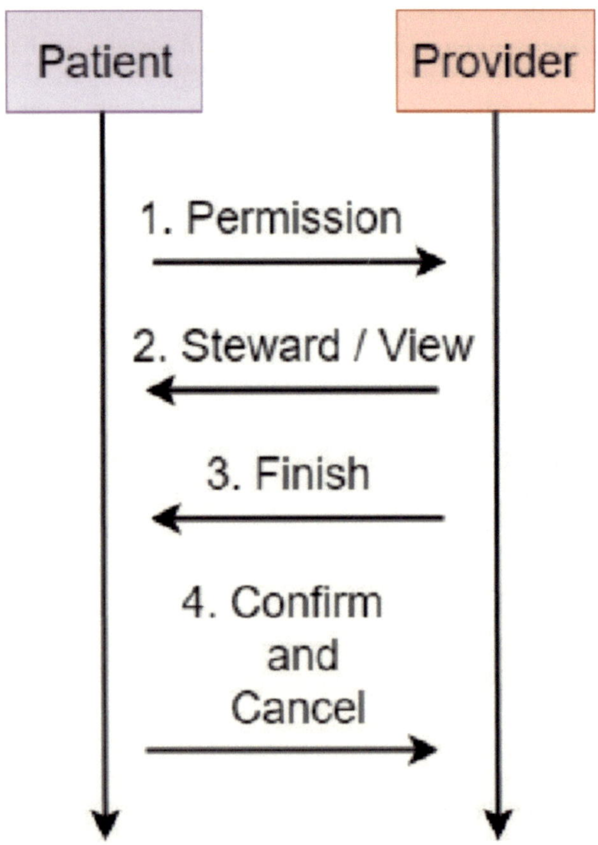


**Figure 2.**
*Authorization procedure of getting medical record permission.*

### *3.1.3 Summary contract (SC)*

This contract holds a list of references to PPRs, locating patients' medical record history. The patient-oriented medical record classification structure in the proposed system is designed. Each record is viewed as an PPR smart contract. The proposed medical record structure is shown in **Figure 3.**

## 3.2 Private blockchain network

The deployment of the private blockchain network is illustrated in **Figure 4**, which is applied for the level of care. The main private blockchain network is plotted

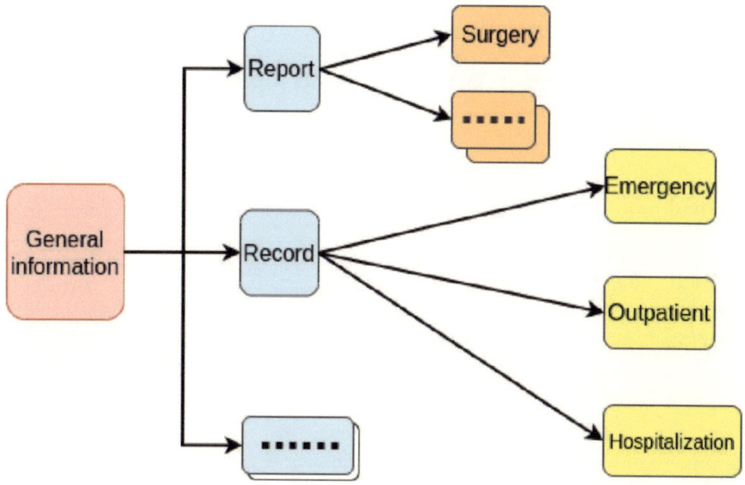


**Figure 3.**
*Proposed personal medical record.*

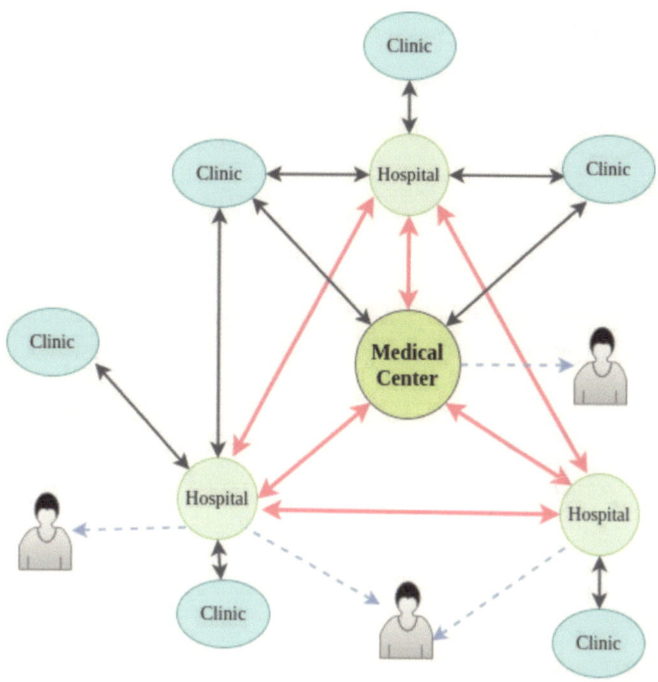


**Figure 4.**
*Deployment of private blockchain.*

by the solid lines. The critical network devices are maintained by the medical centers or hospitals, and the distributed databases among them must be synchronized. The clinics only need to synchronize with the nearby blockchain network nodes to ensure their database stay latest and correct.

Dotted lines stand for the data requests to the blockchain network from patients whom made inquiry for medical record. In this case, the main blockchain network nodes (e.g. the medical center or hospital) are responsible to deal with the requests since their network equipments are capable of handling the heavy network traffic due to plenty of requests. As a primary node in blockchain network, the synchronization speed and correctness should be guaranteed.

### 3.3 System workflow

How the proposed management system works is presented in **Figure 5**.

#### *3.3.1 New entry*

The medical personnel uploads the diagnosis to provider B Node.

#### *3.3.2 External process and outside process*

The external management system detects the updates from the blockchain databases, automatically validates the latest data and notifies the patients of the new updates.

#### *3.3.3 Update nodes*

The blockchain network automatically synchronizes all nodes and offers the latest information to the patient node.

#### *3.3.4 Notification from provider B*

The patient will be notified by the information about who updated the medical record and then checks these updates.

### 3.4 Limitations

With the blockchain-based technique for distributed databases, the additional network facilities and storage devices for network nodes are required to stabilize the whole system. However, it helps save human resource, reduce human errors and accelerate administration process.

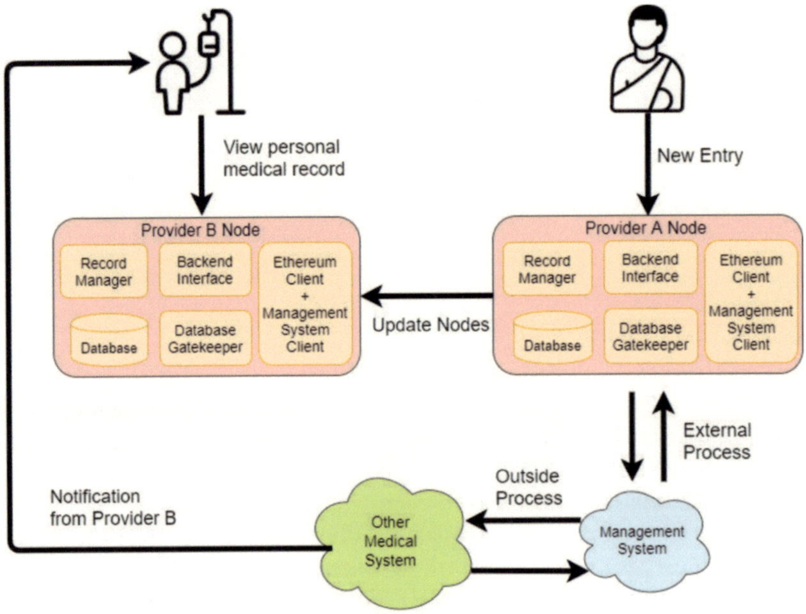


**Figure 5.**
*System workflow.*

| **Research work** | **MedRec** [2] | **MedRec++** [3] | **BBDS** [4] | **MeDShare** [5] | **Liang** *et al.* [6] | **Patientory** [7] | **BSPP** [14] | **Proposed** |
|---|---|---|---|---|---|---|---|---|
| **Identity management** | v | v | v | v | v | v | v | v |
| **Data authenticity** | v | v | v | v | v | v | v | v |
| **Data encryption** | — | v | — | — | — | v | v | v |
| **Blockchain type** | Public | Public | Consortium | Consortium | Consortium | Consortium | Hybrid | Public |
| **Smart contract** | v | v | — | v | — | v | — | v |
| **Data storage** | Off-chain | Off-chain | Cloud | Cloud | Off-chain | Off-chain | Off-chain | Cloud |

**Table 1.**
*Comparison with frameworks.*

| Properties | MedChain [8] | MedRec [2] | MeDShare [5] | Amofa *et al.* [15] | Ancile [16] | FHIRChain [17] | Proposed |
|---|---|---|---|---|---|---|---|
| Access control | v | v | v | v | v | v | v |
| Privacy preservation | v | v | v | v | v | v | v |
| Healthcare legislation | v | v | v | v | v | v | v |
| Smart contracts | v | v | v | v | v | v | v |
| Timed-based smart contracts | v | — | — | v | — | — | v |
| Notification based smart contract | v | v | — | — | v | v | v |

**Table 2.**
*Comparison with other frameworks.*

## 3.5 Comparison with other frameworks

**Table 1** compares the proposed medical record management framework with the existing architectures based on blockchain techniques. Security metrics (identification, data authenticity, data encryption), architecture metrics (blockchain type, data storage) and functionality metrics (smart contract) are all taken into consideration.

**Table 2** also surveys the existing systems. The proposed framework verifies legitimate users before performing the registration process, check the identity to ensure that personal information is only given to authorized users.

# 4. Conclusions

This chapter investigated the blockchain-based and patient-oriented medical record system with the smart contract on EVM. In the future, the drug pedigree may be included into the blockchain. The drug traceability is thus carried out for efficient management and control.

# Acknowledgements

This work was supported by Intelligent Medical Center of E-DA Healthcare Group under Grants EDAH-AI-107-010.

## Author details

Hui Li Wang[1], Shao-I Chu[2*], Jiun-Han Yan[2], Yu-Jung Huang[3], I-Yueh Fang[1], Shu Ya Pan[1], Wei-Cheng Lin[4], Chao-Tien Hsu[5], Chih-Lung Hung[6], Tzung-Ching Lin[6] and Te-Tsun Shen[7]

1 Medical Record Section, E-DA Hospital, Kaohsiung, Taiwan

2 Department of Electronic Engineering, National Kaohsiung University of Science and Technology, Kaohsiung, Taiwan

3 Department of Electronic Engineering, I-Shou University, Kaohsiung, Taiwan

4 Medical Research Department, E-DA Hospital, Kaohsiung, Taiwan

5 Department of Pathology, E-DA Hospital, Kaohsiung, Taiwan

6 Administrative Center, E-DA Hospital, Kaohsiung, Taiwan

7 E-DA Healthcare Group Executive Committee, Kaohsiung, Taiwan

*Address all correspondence to: erwinchu@nkust.edu.tw

## References

[1] H. Jin, Y. Luo, P. Li and J. Mathew, "A Review of Secure and Privacy-Preserving Medical Data Sharing," in IEEE Access, vol. 7, pp. 61656-61669, 2019, doi: 10.1109/ACCESS.2019.2916503.

[2] A. Azaria, A. Ekblaw, T. Vieira and A. Lippman, "MedRec: Using Blockchain for Medical Data Access and Permission Management," 2016 2nd International Conference on Open and Big Data (OBD), Vienna, 2016, pp. 25-30, doi: 10.1109/OBD.2016.11.

[3] H. Yang and B. Yang, "A blockchain-based approach to the secure sharing of healthcare data", Proc. Norwegian Inf. Secur. Conf., pp. 1-12, 2017.

[4] Q. Xia, E. B. Sifah, A. Smahi, S. Amofa and X. Zhang, "BBDS: Blockchain-based data sharing for electronic medical records in cloud environments", Information, vol. 8, no. 2, pp. 44, 2017.

[5] Q. Xia, E. B. Sifah, K. O. Asamoah, J. Gao, X. Du and M. Guizani, "MeDShare: Trust-less medical data sharing among cloud service providers via blockchain", IEEE Access, vol. 5, pp. 14757-14767, 2017.

[6] X. Liang, J. Zhao, S. Shetty, J. Liu and D. Li, "Integrating blockchain for data sharing and collaboration in mobile healthcare applications", Proc. IEEE 28th Annu. Int. Symp. Pers. Indoor Mobile Radio Commun. (PIMRC), pp. 1-5, Oct. 2017.

[7] C. McFarlane, M. Beer, J. Brown and N. Prendergast, Patientory: A Healthcare Peer-to-Peer EMR Storage Network v1, Addison, TX, USA:Entrust, 2017.

[8] E. Daraghmi, Y. Daraghmi and S. Yuan, "MedChain: A Design of Blockchain-Based System for Medical Records Access and Permissions Management," in IEEE Access, vol. 7, pp. 164595-164613, 2019, doi: 10.1109/ACCESS.2019.2952942.

[9] R. Guo, H. Shi, Q. Zhao and D. Zheng, "Secure Attribute-Based Signature Scheme With Multiple Authorities for Blockchain in Electronic Health Records Systems," in IEEE Access, vol. 6, pp. 11676-11686, 2018, doi: 10.1109/ACCESS.2018.2801266.

[10] W. Liu, Q. Yu, Z. Li, Z. Li, Y. Su and J. Zhou, "A Blockchain-Based System for Anti-Fraud of Healthcare Insurance," 2019 IEEE 5th International Conference on Computer and Communications (ICCC), Chengdu, China, 2019, pp. 1264-1268, doi: 10.1109/ICCC47050.2019.9064274.

[11] S. Hasavari and Y. T. Song, "A Secure and Scalable Data Source for Emergency Medical Care using Blockchain Technology," 2019 IEEE 17th International Conference on Software Engineering Research, Management and Applications (SERA), Honolulu, HI, USA, 2019, pp. 71-75, doi: 10.1109/SERA.2019.8886792.

[12] H. Guo, W. Li, M. Nejad and C. Shen, "Access Control for Electronic Health Records with Hybrid Blockchain-Edge Architecture," 2019 IEEE International Conference on Blockchain (Blockchain), Atlanta, GA, USA, 2019, pp. 44-51, doi: 10.1109/Blockchain.2019.00015.

[13] N. Nchinda and A. Cameron, "MedRec: A Network for Personal Information Distribution", International Conference on Computing, Networking and Communications (ICNC), pp. 637-641, 2019.

[14] A. Zhang and X. Lin, "Towards secure and privacy-preserving data sharing in e-health systems via

consortium blockchain", J. Med. Syst., vol. 42, no. 8, pp. 140, 2018.

[15] S. Amofa, E. B. Sifah, K. O.-B. O. Agyekum, S. Abla, Q. Xia, J. C. Gee, et al., "A blockchain-based architecture framework for secure sharing of personal health data", Proc. IEEE 20th Int. Conf. e-Health Netw. Appl. Services (Healthcom), pp. 1-6, Sep. 2018.

[16] G. G. Dagher, J. Mohler, M. Milojkovic and P. B. Marella, "Ancile: Privacy-preserving framework for access control and interoperability of electronic health records using blockchain technology", Sustain. Cities Soc., vol. 39, pp. 283-297, May 2018.

[17] P. Zhang, J. White, D. C. Schmidt, G. Lenz and S. T. Rosenbloom, "FHIRChain: Applying blockchain to securely and scalably share clinical data", Comput. Struct. Biotechnol. J., vol. 16, pp. 267-278, Jul. 2018.